Physiology of l

T. B. MEPHAM

University of Nottingham

Open University Press

Milton Keynes · Philadelphia

Open University Press
Open University Educational Enterprises Limited
12 Cofferidge Close
Stony Stratford
Milton Keynes MK11 1BY, England
and
242 Cherry Street
Philadelphia, PA 19106, USA

First Published 1987

Copyright © 1987 T. B. Mepham

All rights reserved. No part of this work may be reproduced in any form by mimeograph or by any other means, without permission in writing from the Publisher.

British Library Cataloguing in Publication Data
Mepham, T. B.
Physiology of lactation.
1. Lactation
I. Title
612'.664 QP246

ISBN 0–335–15152–3

ISBN 0–335–15151–5 Pbk

Library of Congress Cataloging in Publication Data
Main entry under title:
Mepham, T. B.
Physiology of lactation.
1. Lactation. I. Title. [DNLM: 1. Lactation.
WP 825 M551p]
QP246.M56 1986 599'.036 86–8504

ISBN 0–335–15152–3

ISBN 0–335–15151–5 (pbk.)

Cover design: Schematic representation of three adjacent mammary secretory cells, showing sequential stages (from left to right) in the secretion of fat globules (yellow) and casein micelles (white): see Chapter 6.

Text design by Clarke Williams
Printed and bound in Great Britain by
Butler & Tanner Ltd, Frome and London

Contents

CHAPTER 2: THE STRUCTURE OF MAMMARY GLANDS

CHAPTER 3: THE COMPOSITION OF MILKS

CHAPTER 4: AN ANALYSIS OF LACTATION AS A PRODUCTIVE SYSTEM

CHAPTER 5: BIOSYNTHESIS OF MILK

CHAPTER 6: SECRETORY PROCESSES

CHAPTER 7: ENDOCRINE CONTROL OF MAMMARY GROWTH AND FUNCTION

CHAPTER 8: NEUROENDOCRINE CONTROL OF LACTATION

CHAPTER 9: CONTROL OF MAMMARY FUNCTION AT THE CELLULAR LEVEL

CHAPTER 10: MATERNAL PHYSIOLOGY DURING LACTATION

CHAPTER 11: PROBLEM-SOLVING STRATEGIES

Preface

It is widely held that hardly anyone reads prefaces. Whether this is because experience suggests that they are usually of no consequence or because authors, fearing that they will not be read, *make* them inconsequential, is open to question. Either way it is regrettable. As, on arrival at the theatre or concert hall, the programme notes prepare us for what is coming, so the preface to a book *should* set the tone, summarize the objectives and whet the appetite. That, in any event, is the intention here.

All authors, when they sit down to write, must have a potential readership in mind. In this case, the style and scope of the book were largely influenced by my experience of teaching undergraduate students of animal physiology and zoology, over a number of years; but, since I suspect they represent a fair cross-section of biology students, the book should prove useful to students of biological science in general. Doubtless, because of the practical applications of the subject, it will be seen as most suitable for medical, veterinary and agricultural science students. More generally, however, it might also serve as an introduction, for those of suitable background, to a subject which is increasingly becoming recognized as one of considerable importance to human health and welfare.

The 'physiology' of the title is used in its broader sense, of 'the sum of the physicochemical activities of living organisms': it thus encompasses the separate disciplines of biochemistry, biophysics, endocrinology, neurophysiology, etc. which have emerged from the earlier 'anatomy and physiology' parentage. There is a real sense in which physiology has ceased to be a research discipline, since almost all modern physiological research could be comfortably assigned to one or other of the newer disciplines. But its importance, rather than being diminished, is *enhanced* by this divergence of research activity. More than ever, would there seem to be a need to relate these different modes of enquiry; to attempt to integrate their respective interpretations of biological activity. A major objective of the book is to emphasize the need for such an interdisciplinary approach.

Various devices have been employed in the book to achieve this objective. For example, cross-referencing is liberal; and, at various points in the text, a brief outline of the topic to be discussed is provided, the details of which are filled in later. Two chapters, in particular, take a synoptic viewpoint: Chapter 4, in which lactation is analysed as a 'productive' process, and Chapter 11, in which are discussed practical problems and their possible solutions. The overall structure of the book is summarized in the Venn diagram (Figure 1). Close inspection reveals

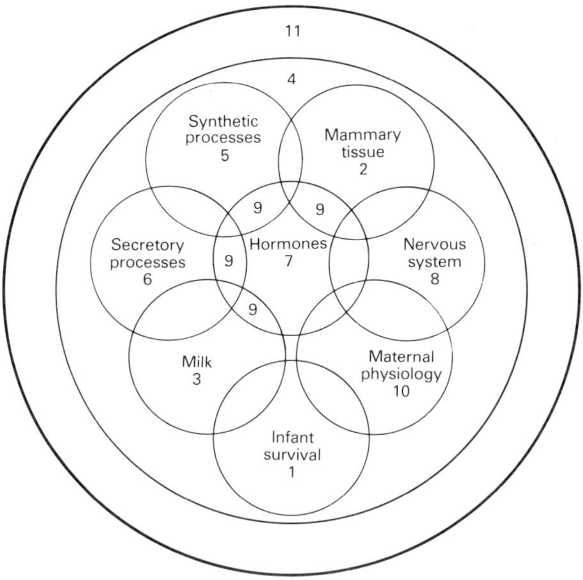

Figure 1. Venn diagram indicating the relationship between chapters of the book. Numerals refer to chapter numbers: Chapter 9 discusses the actions of hormones at the cellular level.

it to be more than simply a pleasingly symmetrical pattern. It may well prove useful to refer back to this diagram when reading the book, to see how the material of the different chapters interrelates.

A criticism of much science teaching is that it places too high a priority on 'information' transfer, and thereby engenders an uncritical attitude on the part of students. Students feel that there are so many facts which need to be learnt that there is no time for questioning, no room for scepticism. Yet the distinctive feature of scientific 'knowledge' is its impermanence: few practising scientists will not have had to modify their opinions quite substantially over, say, the last ten years. In short, science is not simply a body of 'facts', but an *activity*, advances in which are subject to changing ways of thinking, and evolving attitudes as to its utility. Some appreciation of the philosophical and sociological bases of scientific theory would thus seem to be an essential ingredient of the training of any serious student of science. (The writings of T. S. Kuhn provide a stimulating insight into this field of enquiry.) In this book, I have tried to indicate the transitory nature of scientific ideas by frequent allusion to earlier views. While we may derive a certain amusement from the apparent naïveté of our predecessors, we should forbear derision: our own views will doubtless suffer a similar fate.

Rather, then, than the book being seen simply as a 'body of facts', to be mugged-up for examinations, it is also intended, in the late C. H. Waddington's phrase, as *'a tool for thinking with'*.

The perceived uses of scientific knowledge are a major determinant of the type of problems scientists seek to solve and of the means made available for their solution. Virtually all research on lactation nowadays is carried out by scientists working in the *applied* sciences, that is in the fields of medicine and agricultural and veterinary science. But this does not mean, as is commonly assumed, that fundamental physiological considerations are unimportant. Indeed, applied studies which fall into the category of *strategic research* are no less concerned with fundamental processes than are those classed as *pure*. And as a matter of historical fact, many fundamental discoveries have been made by scientists pursuing practical objectives. I take the view that science should be seen *primarily* as a problem-solving activity, a view which is explored more fully in Chapter 11.

Whatever views we might hold about the nature of scientific research, there seems little doubt that biology has recently entered a critical and highly significant phase. The advent of genetic engineering seems likely to place at our disposal extremely powerful tools with which to manipulate biological systems. Whereas, in the past, many people had reservations about the adoption of 'unnatural practices' in medicine and agriculture, this newly acquired ability to tinker with genetic material means that we shall soon be able to change the nature of living organisms 'to order'. In the present context, for example, it might well be deemed more efficient to produce 'milk' *in vitro*, by modifying the genes of bacteria.

Such possibilities are as awe-inspiring as they are exciting. Their future implications demand serious debate (see, for example, Yanchinski, 1985). But they in no way diminish the need for full understanding of the physiological processes by which genetic potential is expressed. The new generation of scientists and technologists will require no less understanding, and perhaps even more wisdom, in the exercise of their skills than has ever been required hitherto.

Ben Mepham
Sutton Bonington

References

Kuhn, T. S. (1977). *The Essential Tension*. (Chicago: University of Chicago Press.)
Waddington, C. H. (1977). *Tools for Thought*. (St Albans: Granada.)
Yanchinski, S. (1985). *Setting Genes to Work*. (New York: Viking.)

Acknowledgements

I am indebted to a number of people for advice and help received while writing the book. In particular, Drs C. H. Knight, N. J. Kuhn, A. J. McArthur, I. H. Mather and G. H. Smith made helpful comments on certain chapters or sections, and I am especially grateful to Dr A. T. Cowie, who read the whole draft manuscript and made several valuable suggestions. Pamela Blythe showed her dependable typographical and telepathic skills in translating my pencilled scrawl into handsome script, a smaller part of this task being performed by Lynne Moseley. Rose Dixon of the Open University Press lent a kindly ear and gave encouraging advice at all the right times. Lastly, I thank my family for enduring with equanimity the long hours I have spent closeted in the study.

Illustrative material was adapted from or inspired by the following sources to whom grateful acknowledgement is also made: Figure 1.3: D. Raphael, *The Tender Gift: Breast Feeding*, Schocken books (1976); Figures 2.3, 2.4, 2.5: A. T. Cowie in *Hormonal Control of Reproduction* 1st edition (ed. C. R. Austin and R. V. Short) Cambridge (1972); Figure 3.1: D. G. Schmidt in *Developments in Dairy Chemistry* vol 1 (ed. P. F. Fox) Applied Science (1982); Figure 3.2: J. E. Butler in *Lactation* Vol 3 (eds B. L. Larson and V. R. Smith) Academic Press (1974); Figure 3.3: L. Hambraeus in *Developments in Dairy Chemistry* Vol. 1 (ed. P. F. Fox) Applied Science (1982); Figure 3.5: O. T. Oftedal *Symp. Zool. Soc. Lond.* 51, (1984); Figure 3.6: R. D. Martin *Symp. Zool. Soc. Lond.* 51 (1984); Figure 3.7: A. T. Cowie *J. Endocrinol.* 44 (1969); Figure 4.7: J. L. Linzell and M. Peaker *J. Physiol.* 216 (1971); Figure 5.2: S. R. Davis and T. B. Mepham *J. Physiol.* 222 (1972); Figure 5.3: T. B. Mepham, P. Gaye and J. C. Mercier in *Developments in Dairy Chemistry* Vol. 1 (ed. P. F. Fox) Applied Science (1982); Figure 5.4: J. C. Mercier and P. Gaye *J. Dairy Sci.* 65 (1982); Figure 5.5: N. J. Kuhn in *Biochemistry of Lactation* (ed. T. B. Mepham) Elsevier (1983); Figure 5.6: M. I. Gurr in *Developments in Dairy Chemistry* Vol. 2 (ed. P. F. Fox) Applied Science (1983); Figure 5.8: E. C. Conn and P. K. Stumpf *Outlines of Biochemistry* 4th edition Wiley (1976); Figure 6.1: J. L. Linzell and M. Peaker *Physiol. Rev.* 51 (1971); Figure 6.3: T. W. Keenan and D. P. Dylewski *J. Dairy Sci.* 68 (1985); Figure 6.4: J. C. Mercier and P. Gaye *J. Dairy Sci.* 65 (1982);

Figures 6.5, 6.6 and 6.7: M. Peaker in *Lactation* Vol 4 (ed. B. L. Larson) Academic Press (1978); Figure 6.9: I. H. Mather and T. W. Keenan in *Biochemistry of Lactation* (ed. T. B. Mepham) Elsevier (1983); Figure 7.1: C. H. Knight *Symp. Zool. Soc. Lond.* 51 (1984); Figure 7.2: A. T. Cowie in *Lactation* (ed. I. R. Falconer) Butterworths (1971); Figure 7.3: T. J. Hayden, C. R. Thomas, S. V. Smith and I. A. Forsyth *J. Endocrinol.* 86 (1980); Figure 7.5: I. R. Fleet, J. A. Goode, M. H. Hamon, M. S. Laurie, J. L. Linzell and M. Peaker *J. Physiol.* 251 (1975); Figure 7.6: A. J. Davis, I. R. Fleet, J. A. Goode, M. H. Hamon, F. M. Maule-Walker and M. Peaker *J. Physiol.* 288 (1979); Figure 7.8: A. T. Cowie in *Lactogenesis* (ed. M. Reynolds and S. J. Folley) University of Pennsylvania Press (1969); Figure 7.9: J. C. Taylor and M. Peaker *J. Endocrinol.* 67 (1975); Figure 8.2: R. Eckert and D. Randall *Animal Physiology* 2nd edition, Freeman (1983); Figure 8.5: I. C. Hart and J. L. Linzell *J. Endocrinol.* 72 (1977); Figure 9.1: E. S. Mills and Y. J. Topper *J. Cell. Biol.* 44 (1970); Figure 9.2: T. B. Mepham, P. Gaye and J. C. Mercier in *Developments in Dairy Chemistry* Vol. 1 (ed. P. F. Fox) Applied Science (1982); Figure 9.3: P. A. Kelly, J. Djiane, M. Katon, M. H. Ferland, L. M. Houdebine, B. Teysott and I. Dusanter-Fourt *Rec. Prog. Horm. Res.* 40 (1984); Figure 9.4: T. Oka in *Biochemistry of Lactation* (ed. T. B. Mepham) Elsevier (1983); Figure 9.5: R. W. Turkington, K. Brew, T. C. Vanaman and R. L. Hill *J. Biol. Chem.* 243 (1968); Figure 9.6: A. J. Davis, I. R. Fleet, J. A. Goode, M. H. Hamon, F. M. Maule-Walker and M. Peaker *J. Physiol.* 288 (1979); Figure 9.7: R. F. Loizzi *Endocrinology* 116 (1985); Figure 9.8: C. H. Knight, A. H. Docherty and M. Peaker *J. Dairy Res.* 51 (1984); Figure 10.1: T. B. Mepham in *Biochemistry of Lactation* (ed. T. B. Mepham) Elsevier (1983); Figure 10.4: E. F. Annison and J. L. Linzell *J. Physiol.* 175 (1964); Figure 10.5: D. E. Bauman and J. M. Elliot in *Biochemistry of Lactation* (ed. T. B. Mepham) Elsevier (1983); Figures 10.6 and 10.8: A. Hanwell and J. L. Linzell *J. Physiol.* 233 (1973); Figure 10.9: R. L. Kleinman and P. Senanayake (eds) *Breast Feeding* IPPF (1984); Figure 11.1 and Table 11.1: UNICEF *State of the World's Children in 1985* Oxford (1985); Figure 11.2: G. J. Ebrahim *Breast Feeding*, Macmillan (1978); Figure 11.3: S. J. Planck and M. L. Milanesi *Bull. World Health Org.* 48, (1973); Figure 11.4: R. L. Kleinman and P. Senanayake (eds) *Breast Feeding* IPPF (1984); Figure 11.7: F. E. Hytten and I. Leitch *The Physiology of Human Pregnancy*, Blackwell (1971); Figure 11.9: P. J. Eppard and D. E. Bauman *Proc. Feed Manuf. Conf.* Cornell University, New York (1984). Tables 3.1, 3.2, 3.5 and 3.6: D. T. Davies, C. Holt and W. W. Christie in *Biochemistry of Lactation* (ed. T. B. Mepham) Elsevier (1983); Table 3.3: R. Jenness in *Developments in Dairy Chemistry* Vol. 1 (ed. P. F. Fox) Applied Science (1982); Table 3.4: L. Hambraeus in *Developments in Dairy Chemistry* Vol. 1 (ed. P. F. Fox) Applied Science (1982).

Abbreviations

The following is a list of abbreviations used in the book.

ACP	acyl carrier protein (of FAS)
ACTH	adrenocorticotrophic hormone
ADH	antidiuretic hormone (vasopressin)
ADP	adenosine 5′-pyrophosphate
ATP	adenosine 5′-triphosphate
AV	arteriovenous (concentration difference)
cAMP	adenosine 3′:5′-phosphate
cGMP	guanosine 3′:5′-phosphate
CNS	central nervous system
CoA	coenzyme A
CT	calcitonin
CTP	cytidine 5′-triphosphate
DHKPGF$_{2\alpha}$	dihyroketo prostaglandin F$_{2\alpha}$
DNA	deoxyribonucleic acid
DNA$_t$	total amount of DNA
EAA	essential amino acids
ECF	extracellular fluid
EM	electron microscope/micrograph
ER	endoplasmic reticulum
ER + (−)	oestrogen receptors present (absent)
FAS	fatty acid synthetase
FFA	free fatty acids
FSH	follicle stimulating hormone
GH	growth hormone
GnRH	gonadotrophin releasing hormone
GSH	glutathoine
GTP	guanosine 5′-triphosphate
HDL	high density lipoproteins
ICDH	isocitrate dehydrogenase shuttle
ICF	intracellular fluid
ICP	insulin, cortisol and prolactin
Ig	immunoglobulin
IGF	insulin-like growth factor
IHD	ischaemic heart disease
ISF	interstitial fluid
J	joule
k(prefix)	kilo ($10^3 \times$)
K_m	Michaelis constant
LDL	low density lipo-proteins
LH	luteinizing hormone
m(prefix)	milli ($10^{-3} \times$)
M(prefix)	mega ($10^6 \times$)

MBF	mammary blood flow (rate)
MFGM	milk fat globule membrane
mol	gram molecule
mRNA	messenger RNA
n(prefix)	nano ($10^{-9}\times$)
NAD(+)	nicotinamide-adenine dinucleotide (oxidized)
NADP(+)	nicotinamide-adenine dinucleotide phosphate (oxidized)
NADPH	nicotinamide-adeninedinucleotide phosphate, reduced
NDP	nucleoside diphosphatase
NEAA	non-essential amino acids
ODC	ornithine decarboxylase
PD	potential difference
PG	prostaglandin
pH	hydrogen ion concentration, minus logarithm of
Pi	orthophosphate (inorganic)
PIF	prolactin inhibiting factor
PPi	pyrophosphate (inorganic)
PR + (−)	progesterone receptors present (absent)
PRF	prolactin releasing factor
PTH	parathyroid hormone
PVN	paraventricular nucleus
RER	rough endoplasmic reticulum
RNA	ribonucleic acid
rRNA	ribosomal RNA
SAMDC	S-adenosylmethionine decarboxylase
SIg	secretory immunoglobin
SON	supraoptic nucleus
SRA	specific radioactivity
TCA	tricarboxylic acid (cycle)
TG	triacylglycerol
TRH	thyrotrophin releasing hormone
tRNA	transfer RNA
UDP	uridine 5′-pyrophosphate
UTP	uridine 5′-triphosphate
VIP	vasoactive intestinal peptide
VLDL	very low density lipoproteins
V_{max}	maximal rate of product synthesis
μ(prefix)	micro ($10^{-6}\times$)

Amino acids

ala	alanine
arg	arginine
asn	asparagine
asp	aspartic acid
cit	citrulline
cys	cysteine
gln	glutamine
glu	glutamic acid
gly	glycine
his	histidine
hyp	hydroxyproline
ile	isoleucine
leu	leucine
lys	lysine
met	methionine
orn	ornithine
phe	phenylalanine
pro	proline
ser	serine
trp	tryptophan
tyr	tyrosine
val	valine

1 Biological Functions of Lactation

1.1. Introduction

Thou hast poured me out as milk
And curdled me as cheese
Job X, 10

The word 'milk' has long had connotations of beneficence and wholesomeness; witness such expressions as, 'the milk of human kindness' and 'a land flowing with milk and honey'. And the human breast similarly has been eulogized in art and literature from earliest times. Nothing, it seems, so aptly portrays our ideals of maternal love and tenderness as the image of a nursing mother and her child.

An experience which was at the same time so common and yet so invested with reverence and significance almost inevitably acquired a folklore rooted in mysticism. Thus Job, in the above quotation, is represented as expressing a notion that was revived in the nineteenth century as the 'globulist theory', according to which body tissues are a coagulum of ingested fluids. Other theories attributed to milk the capacity to bestow on the consumer the temperamental characteristics of the woman or animal from which it originated. Perhaps more than in many areas of biology, progress in the scientific description of lactation has had to counter such 'commonsense' views, which seemed as unassailable as they were familiar.

But with the rise of biochemistry as a discipline in the early years of this century, and the attendant inclination to regard living organisms as 'chemical machines', lactation came to be viewed as an essentially mechanical process. In the 'progressive' West, women who breastfed were labelled 'sentimental cranks'; after all, artificial milk dispensed from a bottle with a rubber teat was so much more *convenient*! Many now consider that this excessively 'reductionist' approach detracts from a full explanation of biological phenomena.

Fortunately, in recent years there has been a growing appreciation of the fact that lactation is not simply a matter of providing adequate nutrients to the newborn. Indeed, in 1976, R.V. Short entitled an article: 'Lactation—the central control of reproduction'. It is the purpose of this chapter to establish the crucial role of lactation as an element in the reproductive strategy of mammals.

1.2. Definition of mammals

The production of milk by the female parent following parturition (or in some cases hatching of the young) is the sole feature by which mammals are distinguished from other vertebrates; while most mammals exhibit viviparity (see section 1.6.2) and possess hairy coats, neither of these characteristics is definitive.

This strategy of feeding the young by secretion of a nutritive product contrasts with that adopted by most lower animals, in which the vast majority of the large numbers of young produced fail to survive. But the strategy is not unique to mammals, being practised by a variety of non-mammalian species, including insects and fish. There are some fascinating parallels with lactation in two species in particular, namely the Discus fish and the pigeon, in both of which the secretion (from the body surface and crop sac, respectively) is sensitive to the same hormone, *prolactin*, which has been found to be so important in the control of lactation in mammals. It is, however, only in mammals that milk is secreted by specific cutaneous organs, the *mammary glands*.

1.3. Significance of lactation in mammalian evolution

It is estimated that there are more than 4000 different species of mammal in existence, although, regrettably, this number is declining. Very largely as a result of the activity of man, they have had an unprecedented effect on the biosphere as a whole.

The radiation of mammals during the transition from the Cretaceous to the Tertiary period (approximately 70 to 150 million years ago) coincided with climatic and vegetational changes which led to the disappearence of dinosaurs and other large reptiles. There has been much speculation as to the precise causes of these changes, but it now appears likely that lactation played a crucial role in enabling mammals to adapt much more readily than reptiles to the new conditions.

One important aspect of lactation is the freedom conferred on the young from the necessity to forage for food. For mammals, unlike reptiles, the habitat need only supply food on which the *adult* can survive and the young may continue to develop in an 'embryonic' fashion instead of diverting their energy to the task of survival. Possibly in consequence, mammals reach sexual maturity much earlier than reptiles of equivalent body weight. Furthermore, lactation frees the mother from the disadvantage of having to carry the young *in utero* for a prolonged period, a consideration of great importance in species relying on speed to escape predators or on mobility to search for suitable food.

As a means of feeding the young, lactation has two specific advantages over the practice adopted by birds. Firstly, the nutrient supplied to the young can be 'stored' in the form of maternal tissues, from which the precursors of milk may be mobilized as circumstances demand. Secondly, short-term variations in the quantity or quality of food eaten by the mother have relatively little impact on the nutrition of the young, for essentially the same reason. Lactation is also believed to be important in relation to dental development in mammals, since it allows a delay in the appearance of teeth until the skull and jaws are almost fully grown. In consequence only two sets of teeth are required in mammalian development (that is, the condition of *diphyodonty*) in contrast to the numerous sets of teeth involved in reptilian development (*polydiphyodonty*).

Finally, it seems reasonable to ascribe to lactation a crucial role in the development of social organization in mammals. In contrast to reptiles, where nutritional requirements of adults and young are widely different, mammals provide food for the young until they are no longer vulnerable to predation or competition from other adults. This sustained physical proximity between adults and young, in conditions which are initially non-competitive, would seem likely to have provided a significant stimulus to social organization. Similarly, such conditions may have been conducive to 'learning', that is, the acquisition, initially through mimicry, of those skills on which survival of the individual depends. With the development of secure, role-differentiated human societies, the exercise of such skills would appear to have given rise to all those highly valued activities and achievements we now class as 'culture'. Perhaps not without justification, we might perceive the human breast, innocently elaborating a seemingly bland fluid, to be the very font of human civilization.

1.4. Phylogeny of the mammary gland

The functional mammary gland is a compound tubuloalveolar structure lying beneath the skin and exterior to the body cavity. Milk produced in the gland is removed by the suckling young through teats or nipples opening onto the body surface. From which structure in non-mammalian species did the mammary gland evolve? While it is possible that the gland developed *de novo* from ectoderm and underlying mesenchymal tissues, and is therefore homologous to no particular non-mammalian structure, it is thought more probable that it arose by evolutionary modification of one or more of the skin glands which are widely distributed on the surface of mammals. The closest homology appears to be with a type of sweat gland which secretes an opalescent fluid, components of which, in certain species, act as pheromones and as such are important in sexual attraction. In humans, such glands are prevalent in the armpit and in the pubic and anal regions. They exhibit an apocrine mode of secretion (see section 6.1).

Based on this alleged homology, an attractive, though admittedly speculative theory, has been proposed for the origin of the mammary gland. It is reasoned that areas in which mammary glands arise (bilaterally, on the ventral, abdominal and pectoral surfaces) are equivalent to the 'brooding spots' of reptilian ancestors.

These areas would have had a rich blood supply, which with the development of the hairy coat would have favoured the rapid growth of associated skin glands. It is apparent that there was no sudden transformation of reptile into mammal and the therapsid reptiles from which most mammals are believed to have evolved are thought to have had hair.

The emergence of lactation demanded more than the appearance of a milk-secreting organ; it also required instinctive sucking behaviour by the young. In a typically erudite and provocative paper, almost his last publication, J.B.S. Haldane (1965) speculated on the origin of this activity. (In the course of the article, he not only cited Shakespeare and Aristotle in support of his case, but also predicted that future human fathers will receive 'biochemical treatment' to develop maternal instincts!) Haldane surmised that the reptilian ancestors of mammals living in hot, dry climates cooled their eggs by periodically immersing themelves in water. On hatching, the young would have relieved their thirst by sucking at the wet maternal hair, a behavioural response which became instinctive. Thus, according to this theory, sucking behaviour *preceded* the emergence of the mammary gland and was able to exploit it fully when the latter occurred. It is presumed that, through the processes of natural selection, compositional changes in the secretion of the skin glands resulted in what we now call 'milk'.

1.5. Brief outline of lactational physiology

At this point, it is useful to consider in broad outline the principal physiological features of lactation. This treatment will be concise in the extreme and the rest of the book may be regarded as an amplification of this outline and an investigation of its biological implications.

In most species, milk is largely composed of water. It contains hundreds of different compounds, the three quantitatively major chemical classes of which are fats, proteins and carbohydrates. The principal chemicals in these classes, together with some of the minor components, do not occur elsewhere in the body, so that their presence in milk depends on prior synthesis in the cells of the mammary gland from materials delivered in arterial blood. The synthesizing cells are usually referred to as *secretory cells*, or, collectively, as the *secretory* (or *mammary*) *epithelium*. The cells are arranged in single layers lining the hollow lumina of spherical structures called *alveoli* (see Figure 2.3, page 18), but for the present purposes we may simply regard the mammary epithelium as a continuous monolayer separating interstitial fluid (ISF) on one (the blood) side from milk on the other (Figure 1.1).

Milk produced in the secretory cells is transferred from the alveolar lumina to larger ducts and cisterns within the gland, where it is stored in the intervals between periods of removal by the suckling young. Milk removal is facilitated by contractile cells called *myoepithelial cells*, which surround each alveolus; when stimulated to contract, the pressure within the alveolar lumina increases and milk is forcibly ejected into the ducts of the gland, from which it is readily removed by the young.

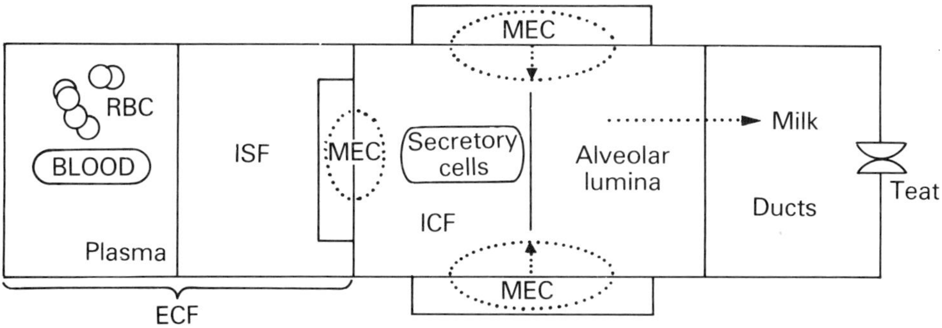

Figure 1.1. Schematic model of the relationship between mammary cells and body fluids. MEC, myoepithelial cells (dotted, contracted); RBC, red blood cells; ISF, interstitial flud; ECF, extracellular fluid; ICF, intracellular fluid.

There is, unfortunately, no consensus on the appropriate use of the term *suckling*, which is variously used to describe the activity of the mother, the young or both. To avoid confusion, the term *nursing* will be employed for the maternal component of the act, with 'suckling' being reserved for the infant's activity. However as discussed later (see section 8.2.2) the notion that suckling activity necessarily involves 'sucking' (that is, creation of a negative pressure) is erroneous.

Myoepithelial cell contraction is brought about by a neuroendocrine reflex involving the relaying of nervous impulses from pressoreceptors in the teats (activated by the suckling young) to the central nervous system, which results in the release of hormones from the pituitary gland (Figure 1.2). *Oxytocin*, from the

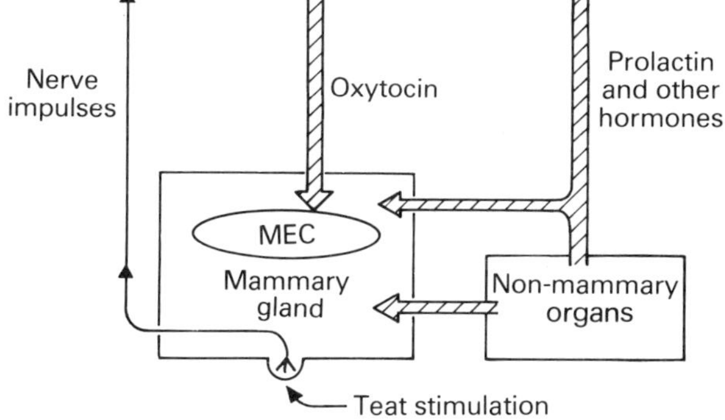

Figure 1.2. Highly simplified scheme showing neuroendocrine reflexes involving the mammary gland. MEC, myoepithelial cells; A, anterior lobe of pituitary gland; P, posterior lobe.

posterior part of the gland, is transported in the blood to the mammary glands, where it elicits myoepithelial cell contraction. Other hormones, like prolactin and *growth hormone* from the anterior part of the gland, are important by virtue of their roles in maintaining the synthetic and secretory capacities of the secretory epithelium. Some of the latter, together with other hormones, have effects on non-mammary organs and tissues which profoundly influence the process of lactation by, for example, regulating the availability in the blood of substances required for milk synthesis, or by depressing fertility.

Since lactation is a phasic process, usually only occurring for a period following birth of the young, the mammary glands exhibit cyclical changes in activity. Mammary growth and development (both at and following puberty and during successive pregnancies) is referred to as *mammogenesis*. The initiation of milk secretion at about the time of parturition is termed *lactogenesis*, while the subsequent maintenance of secretion (a process dependent on continued suckling or artificial milking) is *galactopoiesis*. (The latter term is also applied to augmentation of an established yield, agents or procedures stimulating yield being said to have *galactopoietic effects*.) All three stages are subject to hormonal control, and in the latter case in particular, neural stimuli are also important. The mammary secretion in early lactation (*colostrum*) contains high concentrations of *immunoglobulins*, which, because they carry antibodies, are important in many species in conferring on the newborn resistance to infectious diseases.

1.6. Mammalian superorders

Zoologists divide the class Mammalia into three superorders, the Ornithodelphia, the Marsupialia and the Eutheria. Although the classification is not based on lactational criteria, it is clear from the following that it coincidentally identifies some interesting differences in the form this activity may take.

1.6.1. Ornithodelphia

This contains a single order, the Monotremata, comprising egg-laying mammals, indigenous only in Australasia. There are two species of monotreme, the duck-billed platypus and the spiny anteater. The former usually lays two eggs 15 days after mating, and after the eggs have hatched the young remain in the nest for 15 weeks obtaining milk from mammary glands which open onto the mother's abdomen (Figure 1.3); there are no teats. In the spiny anteater, a pouch develops on the abdomen at the beginning of the breeding season and when the egg is laid it is transferred by the beak into the pouch, where it subsequently hatches. Milk is sucked from abdominal slits which open into the pouch, a process which is atypical of that usually employed by species in which teats are present (see sections 1.5 and 8.2.2).

1.6.2. Marsupialia

Marsupials give birth to live young (*viviparity*) rather than, like monotremes, hatching them from eggs (*oviparity*). Although popularly categorized as pouched-

Figure 1.3. A duckbilled platypus nursing three young. (Modified from a drawing by S. Greco in Raphael, 1976.)

animals, the principal feature which distinguishes them from other mammals is the distinctive relationship between the genital and excretory ducts, and with no true placenta the gestation period is characteristically short (only 3–6 weeks) in comparison with the period of ensuing lactation. Marsupials include species like the opossum, bandicoot, koala bear and kangaroo.

The importance of lactation is indicated by considering the case of the red kangaroo. The newborn ('joey'), weighing only 0.75 g, on reaching the pouch attaches itself to one of the four teats and remains so attached for several months. The teat, which extends well into the pharynx, elongates rapidly in response to suckling. The joey first looks out of the pouch at about 150 days, when it weighs about 1 kg, and by 190 days, when it first leaves the pouch, it weighs 2 kg. However, it continues to enter the pouch for feeding until about 230 days, and even returns for milk up to the age of 1 year, though too large then to enter. Frequently, a second embryo begins to develop just after the first has reached the pouch. However, suckling induces endocrine changes which arrest further development and delay implantation until the first joey begins to be weaned. The subsequent implantation and birth of the second joey mean that the kangaroo often nurses two offspring of vastly differing size and maturity, and supplies each with milk of volume and composition related to its need. The neuroendocrine basis of this effect is discussed in section 10.5.1.

1.6.3. Eutheria

Eutherians, which account for 95% of all mammalian species, are distinguished from other mammals by the development in the pregnant female of a placenta, by means of which the fetus receives nutrients *in utero*. In consequence, although there is much variation between species in the state of maturity of the newborn, they are invariably more mature than newborn marsupials. As in marsupials, teats or nipples are always present.

For the purposes of discussion, it is useful to divide eutherians into a number of 'lactational classes', because there is a very wide divergence in the forms lactation takes. This is hardly surprising in view of the great variety in size, physique, length of gestation, habitat and behavioural characteristics exhibited by different species. For example, some have very immature (*altricial*) young, others precocious young (Figure 1.4). Some are wholly or partially aquatic, some

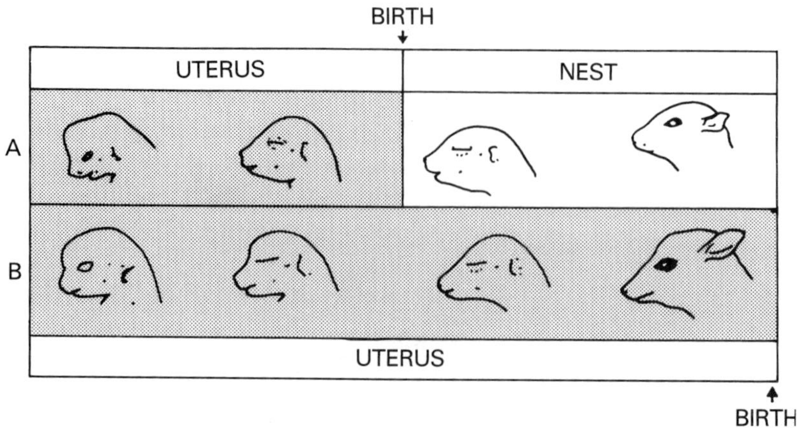

Figure 1.4. The distinction between altricial (A) and precocious (B) mammalian neonates. In precocious mammals the nest phase is, as it were, included in gestation. (Reproduced from Martin, 1984, by permission of Academic Press and R.D. Martin.)

live in polar regions, others in the desert. Since in each case, the newborn is thrust at birth into an alien environment, colder than that in which it spent its previous existence, and is precipitately deprived of its source of food and oxygen, the particular 'lactational strategy' adopted to deal with this desperate situation, and to nurture the young until it is able to lead an independent existence, must necessarily be finely tuned to its new environment. The trauma of birth must for many be surpassed only by that of death. In the following two sections, examples of the two major categories of newborn are discussed.

1.6.3.1. Altricial young In many species of rodent, such as rats and mice, the young are born after a short gestation, and at birth are quite helpless, having closed eyes and ears and no hair or teeth. Heat loss is a significant problem, although it is ameliorated by their short limbs and tails which reduce the 'surface area:volume' ratio. Body heat is maintained by the strong behavioural urge of the mother to keep the litter in very close bodily proximity within the nest and, indeed, the mother spends about 80% of her time nursing. Immature rodents are dependent on the mother not only for food and warmth but also for disposal of excreta. During the first week or so, they are unable to spontaneously empty their bowels or bladders: the mother assists this process by manipulating them with her forepaws to stimulate evacuation, and then consuming the excreta.

Rabbits, like rodents, have very altricial young, but, in contrast, the mother forsakes them for long periods, even up to 16–20 hours at a time, a behavioural pattern possibly evolved to avoid the undue attention of predators. This is possible because milk ejection is a very powerful reflex in rabbits and the milk obtained within a matter of minutes satisfies the young for the next 24 hours. Moreover, body temperature is maintained both by restriction of heat loss by siting the nest underground and by generation of heat in brown adipose tissue.

1.6.3.2. Precocious young Young of artiodactyls (cloven-hoofed species) such as pigs, cows, sheep and goats (the latter three being ruminants) and the perissodactyls, such as horses, zebras and rhinoceroses, all have young which are quite mature at birth (Figure 1.5). The mother usually makes few, if any, preparations

Figure 1.5. A goat nursing its newborn kid, weighing 3.2 kg. (Courtesy, Mr P. M. Taylor.)

for parturition and some, such as giraffes, give birth standing up. The newborn are usually physically well developed and homoiothermic: they can see, hear, walk and, in the case foals, run with exuberance. Milk is, however, of vital importance and in artiodactyls, in particular, the young are highly dependent on the passive immunity acquired through the secretion of immunoglobulins in colostrum.

1.7. Lactation: a crucial element of reproduction

It is apparent that the popular conception of lactation as simply a matter of food supply is inaccurate. Quite apart from the nutritional and immunological properties of milk and colostrum, lactation, through its reliance on physiological and psychological bonding between parent and infant, has far-reaching consequences for the survival of mammalian species. From a purely biological viewpoint, the 'purpose' of life is to generate more life. Reproduction, therefore, is only biologically successful if it results in offspring which can themselves reproduce successfully.

Biologically useful life can thus be regarded as a series of cycles in which an individual has to surmount a number of obstacles. Figure 1.6 illustrates the ways in which lactation, and the behavioural responses it necessitates, aids in overcoming such obstacles, though not all are important in all species. The six properties of lactation identified in Figure 1.6 have all been referred to earlier in this chapter and most will be discussed further in later chapters.

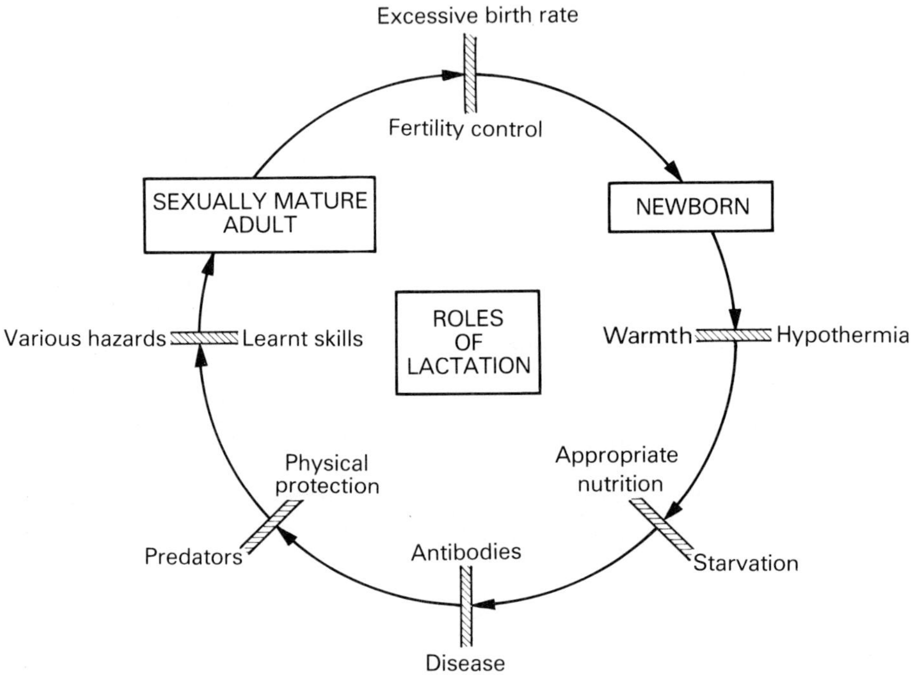

Figure 1.6. The biological roles of lactation. Six barriers to the attainment of sexual maturity are identified (outside the circle) and the lactation-associated factors which allow these barriers to be surmounted are shown inside the circle. The relative importance of the different factors varies with species.

1.8. Breastfeeding in the human

In devoting this section to breastfeeding the author is not merely betraying an anthropocentric bias. More to the point, we, as a species, appear to be the only mammals to have reached the conclusion that lactation is dispensible; that the mechanical dispensation of modified cow's milk through a rubber teat is a satisfactory alternative. Such a 'mechanistic' approach is the inevitable consequence of a reductionist conception of lactation, which regards it as simply a means of nutrient provision. It is clear that the chemical content of milk is but one aspect of the role of lactation in preparing the newborn for independent existence. Granted that in the majority of human societies babies will not be subject to hypothermia or predation, there still remain significant respects in which a bottlefed baby is deprived of important provisions, such as in the attenuation of the psychological bonding between mother and infant, and in the loss of the protective role of colostral immunoglobulins. Moreover, not only are artificial milk powders deficient in many respects in comparison with breast milk, but the demand for sterility in their preparation is not always easily met.

Problems are usually much more acute in poorer countries. D. B. and E. F. P. Jelliffe (1978) have examined the disadvantages of bottlefeeding in relation to resource-rich communities (that is, developed western societies) and resource-poor communities (that is, the third world). They distinguish seven actual or possible disadvantages of bottlefeeding:

nutritional/metabolic consequences (such as infantile obesity; iron-deficiency anaemia)
allergic consequences
sudden infant death syndrome (for example, endotoxaemia)
infectious consequences (for example, septicaemia)
emotional consequences
alimentary consequences (such as dental cares), and
economic consequences (for example, the cost of bottlefeeding is approximately three times that of breastfeeding in the United States).

In resource-poor communities the hazards of bottlefeeding include all those listed, plus

nutritional consequences (for example, kwashiorkor; nutritional marasmus; vitamin deficiency), and
infections (for example, diarrhoeal disease; otitis media).

Quite apart from such consequences, bottlefeeding is believed to exacerbate population growth in those countries in which artificial contraception is not widely practised. In primitive societies like the !Kung (the exclamation mark represents the sound made by clicking the tongue), who live in the Kalahari Desert and demand-feed their babies for several years, the induced lactational anovulation results in an average birth interval of 4 years. But with the adoption of bottlefeeding, this interval is markedly reduced. Thus, not only are babies prey to malnutrition and disease, but they are also born with increasing rapidity.

In contrast to such major effects, some of the other functions of lactation, which relate to mother–infant behavioural interactions, may seem insignificant, though possibly only because we find it difficult to quantitate them. Figure 1.7 summarizes the ways in which the mother and infant interact. The fluid transfer is represented at the bottom of the diagram, but not even the whole of the lower right block would apply if the baby were being bottlefed by a stranger. However, in sympathetic breastfeeding there are clearly many stimuli arising from both individuals, some of which are intentional (like stroking and cooing by the mother)

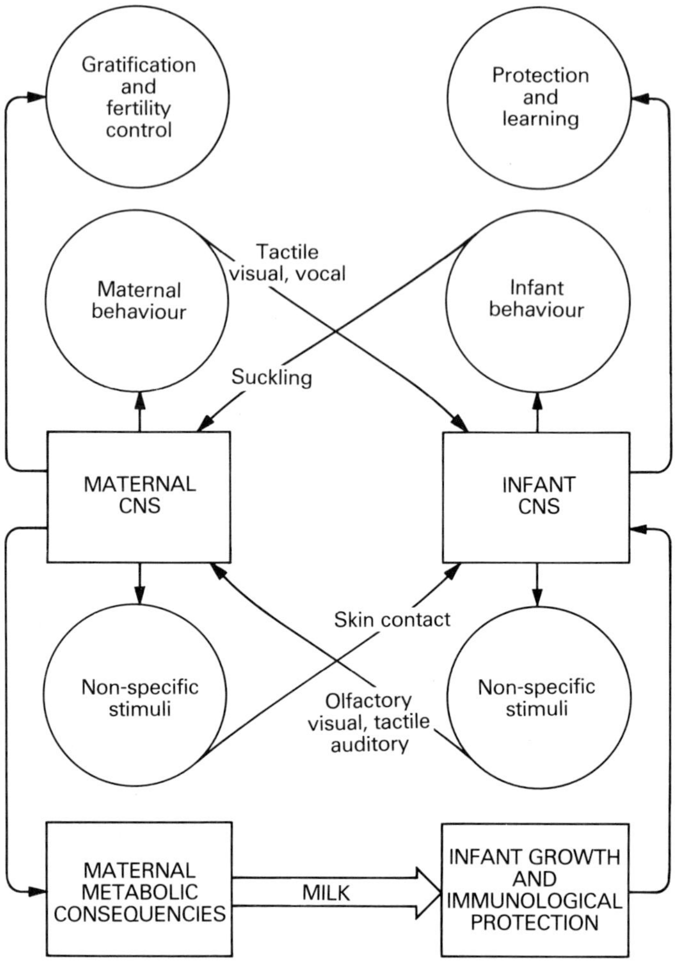

Figure 1.7. The interaction between maternal factors (on the left) and infant factors (on the right) during the act of nursing.

and others incidental, but possibly no less important (like bodily contact and the sound of the mother's heart beat). Under such conditions the baby begins to learn to communicate, initially through mimicry and other forms of imitative behaviour; while the mother derives feelings of warmth and pleasure which might be seen as essential correlates to a form of behaviour which otherwise provides her with no tangible benefits (though lactation is a potent slimming aid).

1.8.1. Protection from breast cancer

A further claim for the importance of lactation in women has been made, in the past, in respect of protection from cancer. Recent aetiological evidence, however, points more clearly to the correlation between 'a young age at first pregnancy' and low incidence of cancer development. The question is not fully resolved, and the observation that Chinese Tanka women (from Hong Kong fishing villages), who customarily nurse only with the right breast, subsequently show a greater incidence of cancer in the left breast at the time of menopause, clearly merits closer study.

1.8.2. The breast in sexual libido

The role of the female breast in sexual arousal of the male seems to be an important element of libido in most societies, although this is apparently a uniquely human characteristic. Desmond Morris, in his popular book *The Naked Ape*, speculated that the breasts and the red lips of women perform a role analogous to the buttocks and labia of other primates. Whether or not this is the case, libido would seem to be a normal prerequisite of fertility.

Natural and artificial contraceptive practices represent ways of dissociating libido from fertility, the necessity for which assumes growing importance as the world's population exceeds the supply of resources needed to support it. An analogy drawn by Havelock Ellis is pertinent. The hand, he pointed out, is an organ the evolution of which initially proved advantageous in food-gathering and tool-making: yet hands are now the agents of expression in art, music and literature. An analogous role-transformation in respect of sexual libido would seem to be imperative for human welfare on the global scale.

1.9. Conclusion

Lactation may be viewed as a form of survival strategy. It is difficult to compare its success with strategies employed by other species, vertebrate and non-vertebrate, because 'success' is capable of different interpretations. But it is obviously the case that mammals possess far more complicated nervous systems than other living organisms and that this has resulted in the evolution, in man, of mental powers which, collectively, are of seemingly limitless potential.

Lactation nicely illustrates the overriding importance of what, in computer terminology, is called 'software', that is, the program encoded in the genes. It is as though, once skin glands had proved capable of adopting a nutritive role, other systems of the body were 'roped-in' to subserve their needs. For example, prolactin has many and diverse functions in fish, amphibia and birds; peptide hormones structurally similar to oxytocin are important in invertebrates; and steroid hormones, similar to those involved in mammogenesis perform a regulatory role in insect moulting. Moreover, α-lactalbumin, a milk protein of paramount importance in the control of milk secretion, appears to be evolutionarily related to lysozymes, a widely dispersed group of enzymes. 'In contrast to the engineer, evolution does not produce innovations from scratch. It works on what already exists, either transforming a system to give it a new function or combining several systems to produce a more complex one' (F. Jacob, 1982).

Further reading

Blaxter, K. L., (1971). The comparative biology of lactation. In *Lactation*. pp. 51–69. ed. Falconer, I. R. (London: Butterworths.)

Haldane, J. B. S. (1965). The possible evolution of lactation. *Zool. Jb. Syst. Bd.* **92**,, 41–8.

Hogarth, P. J. (1976). *Viviparity*. (London: Edward Arnold.)

Ing, R., Ho, J. H. C. and Petrakis, N. L. (1977). Unilateral breast feeding and breast cancer. *Lancet* **ii**, 124–7.

Jacob, F. (1982). *The Possible and the Actual.* (New York: Pantheon Books.)

Jelliffe, D. B. and Jelliffe, E. F. P. (1978). *Human Milk in the Modern World.* (Oxford and New York: Oxford University Press.)

Long, C. A. (1969) The origin and evolution of mammary glands. *Bioscience* **19** (6), 519–23.

Martin, R. D. (1984). Scaling effects and adaptive strategies in mammalian lactation. In *Symp. Zool. Soc. Lond.* **51** 87–117.

May, R. M. and Rubenstein, D. I. (1985). Reproductive strategies. In *Reproductive Fitness* (Vol. 4 of *Reproduction and mammals*) 2nd edition, pp. 1–23. eds Austin, C. R. and Short, R. V. (London and New York: Cambridge University Press.)

Pond, C. M. (1984). Physiological and ecological importance of energy storage in the evolution of lactation: evidence for a common pattern of anatomical organization of adipose tissue in mammals. *Symp. Zool. Soc. Lond.* **51**, 1–32.

Short, R. V. (1976). Lactation—the central control of reproduction. In *Breast-feeding and the Mother* pp. 73–6. Ciba Foundation Symposium 45 (Amsterdam: Elsevier).

Tyndale-Biscoe, H. (1973). *Life of Marsupials.* (London: Edward Arnold.)

Widdowson, E. M. (1981). *Feeding the Newborn Mammal.* (Burlington, N. Carolina: Carolina Bio-

2 The Structure of Mammary Glands

2.1. Introduction

In the last chapter we took an overall view of lactation: its evolution, species variability and role in the reproductive process. Now we need, as it were, to don our lab coats: to consider the morphological structure of the glandular tissue that manufactures milk, and (in Chapter 3) the chemical structure of the milk itself.

Despite marked differences between species in the relative size, number and anatomical position of the glands, their histological structure is remarkably uniform. The discussion will concentrate on the structure of fully lactating glands. Changes occurring during puberty, pregnancy and at involution are discussed in Chapter 7.

2.2. Morphology

2.2.1. Gland number and location

There is much interspecies variation in the number and position of mammary glands: for example, two pectoral glands (in women and elephants), two inguinal glands (in goats and guinea-pigs), four inguinal glands (in cows) and numerous, up to 12, distributed along the whole abdominal surface (in rats, cats and pigs). The record number reported is for the hedgehog-like insectivore, the tenrec, which has 22 to 24 glands, and, while they are present in most species as bilateral pairs, one each side of the ventral midline, some marsupials have an odd number. The coypu is unusual in having glands so high on the abdomen that they occupy a virtually dorsal position (Figure 2.1).

2.2.2. Lobuloalveolar system

In common with all exocrine glands, the parenchyma of mammary glands comprises two types of tissue, *secretory* and *ductal*. The former consists of

Figure 2.1. A female coypu nursing a litter of eight 16-day-old offspring. (Drawing by L. M. Gosling.)
(Reproduced from Gosling, Baker, and Wright, 1984, by permission of Academic Press and L. M. Gosling.)

specialized epithelial cells (*secretory cells*) grouped in numerous pear-shaped *alveoli*, in which the cells are arranged in a single layer lining the hollow lumen. The scanning electron micrograph (Figure 2.2) of rat mammary tissue shows about 15 such alveoli, each of which appears to comprise about 50 secretory cells. Milk is secreted into the hollow lumina of the alveoli, from which it drains via narrow ductules into larger-bore ducts which constitute the *duct system*. The arrangement of alveoli and their contiguous ductules gives the appearance of bunches of grapes on twigs (Figure 2.3), a configuration defined by the term *racemose* (from the Latin for 'a bunch of grapes'); an alternative term for 'alveolus' is *acinus* (from the Latin for 'a grape'). Groups of neighbouring alveoli drained via a common duct constitute a *lobule*, adjacent lobules being separated from each other by connective tissue septa. Some cells of the smaller ductules may also be capable of milk secretion, and these, together with the cells lining the alveoli, are collectively termed the *lobuloalveolar system*.

2.2.3. *Ducts*

In considering the arrangement of large ducts we encounter some prominent interspecies differences (Figure 2.4). For example, in some species (including dairy cows and goats) the large ducts (*galactophores*) empty into voluminous *sinuses* (also called *cisterns*), which in a full gland may accommodate half of the milk it contains. In the cow (Figure 2.4) and goat (Figure 2.5) it is usual to refer to the *gland cistern* and *teat cistern* as separate cavities. In other species (such as rats, guinea-pigs), while the ducts empty into a common reservoir at the base of the

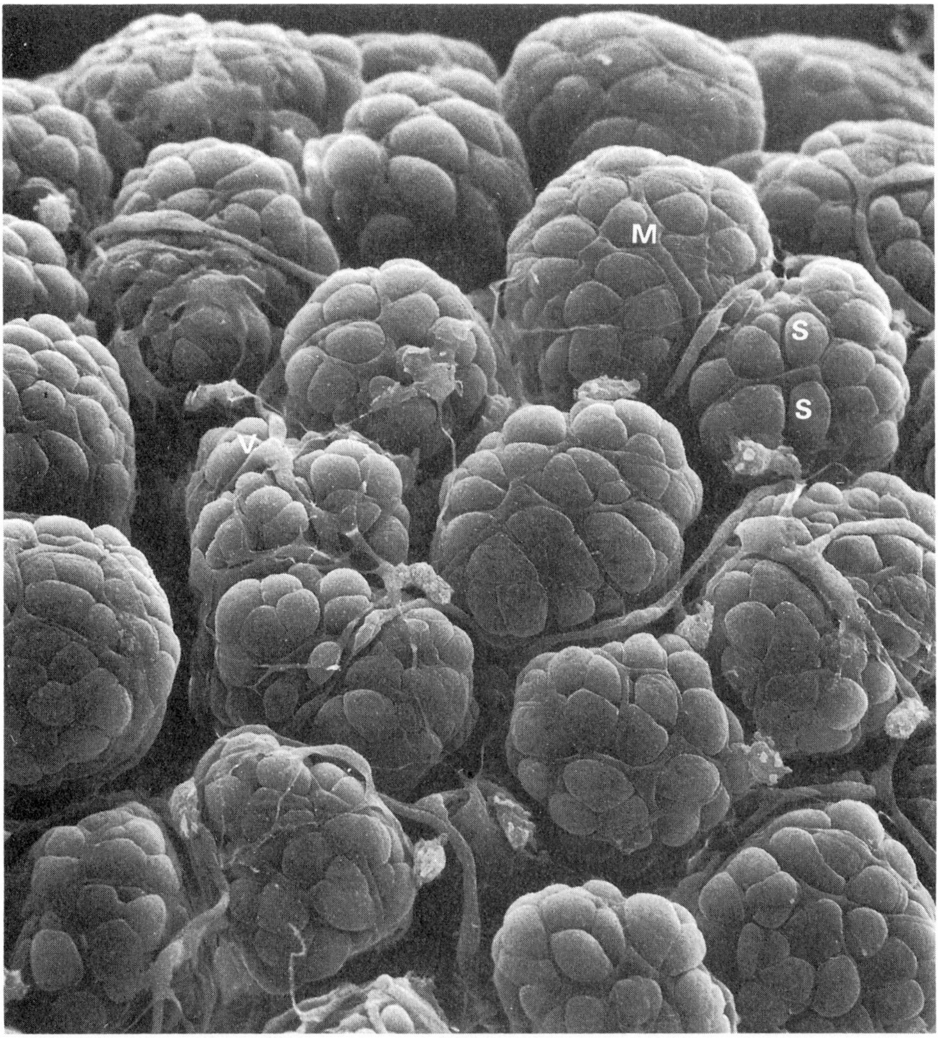

Figure 2.2. Scanning electron micrograph of mammary alveoli from a lactating rat (× 1000). Most connective tissue had been removed by enzyme and acid treatment. V, small blood vessels; M. myoepithelial cells; S, individual secretory cells.
(Reproduced from Nagato *et al.*, 1980, by permission of Springer-Verlag and T. Nagato.)

teat, its capacity is small, so that most milk produced is located in the lobuloalveolar tissue. In both of these types of arrangement milk is removed from the glands via a single orifice in the teat. By contrast, in for instance rabbits and women, each of the ducts draining separate lobules, rather than supplying a common sinus, terminates at the nipple, which is thus perforated by a number of orifices (about 12 in women). Essentially the same arrangement exists in monotremes except that there are no teats and the ducts open onto the body surface at the base of hairs. The nature of the duct system in different species

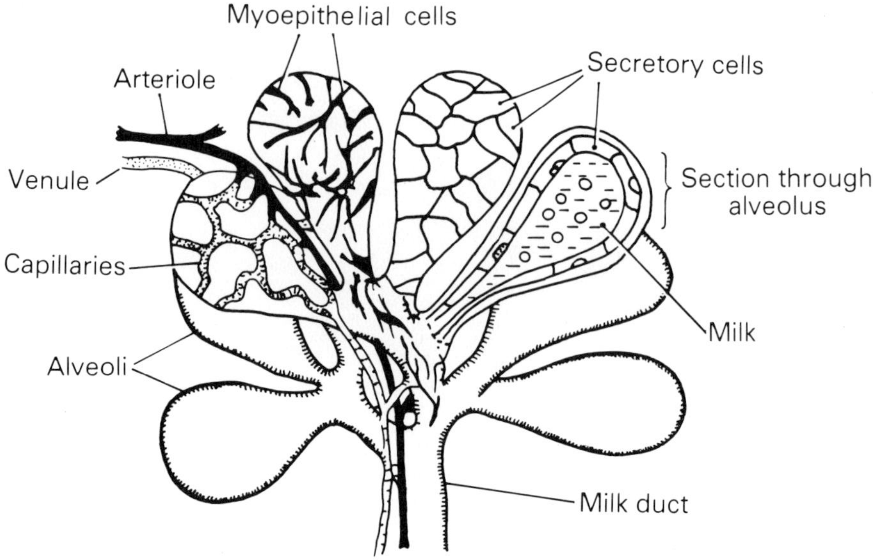

Figure 2.3. Diagrammatic representation of a cluster of mammary alveoli. (Adapted from Cowie, 1972.)

has implications for the mechanisms employed in milk removal (see section 8.2).

2.2.4. Stroma

Those elements of the mammary tissue not included in the parenchyma are collectively called the *stroma,* a term which encompasses the skin, connective tissue, adipose tissue, blood and lymph vessels and nerve tissue.

2.2.5. Blood and lymph vessels

Since mammary glands are skin structures, they share the blood supply to contiguous skin areas. Thus, in animals with pectoral glands the mammary arteries are branches of the thoracic and pectoral arteries, whereas in those with inguinal glands blood is supplied via the external pudic arteries (Figure 2.6). The same considerations apply to venous and lymphatic vessels.

The arterial supply terminates in networks of capillaries surrounding each alveolus (see Figure 2.3). Studies in goats, dogs and cats show that each alveolus is supplied by 5–12 capillaries, the distance between neighbouring vessels being about 10–15 μm. Mammary tissue may also contain *vascular arteriovenous (AV) shunts,* that is, vessels which either directly connect arteries to veins (*arteriovenous anastomoses*) or arterioles to venules (*arteriovenular bridges*), and thereby bypass secretory tissue (Figure 2.7).

It was originally thought that the smallest lymphatic vessels were present in alveoli, but it has now been established that they are confined to lobuloalveolar tissue, so that tissue fluid enters the finest lymphatics which surround the lobules.

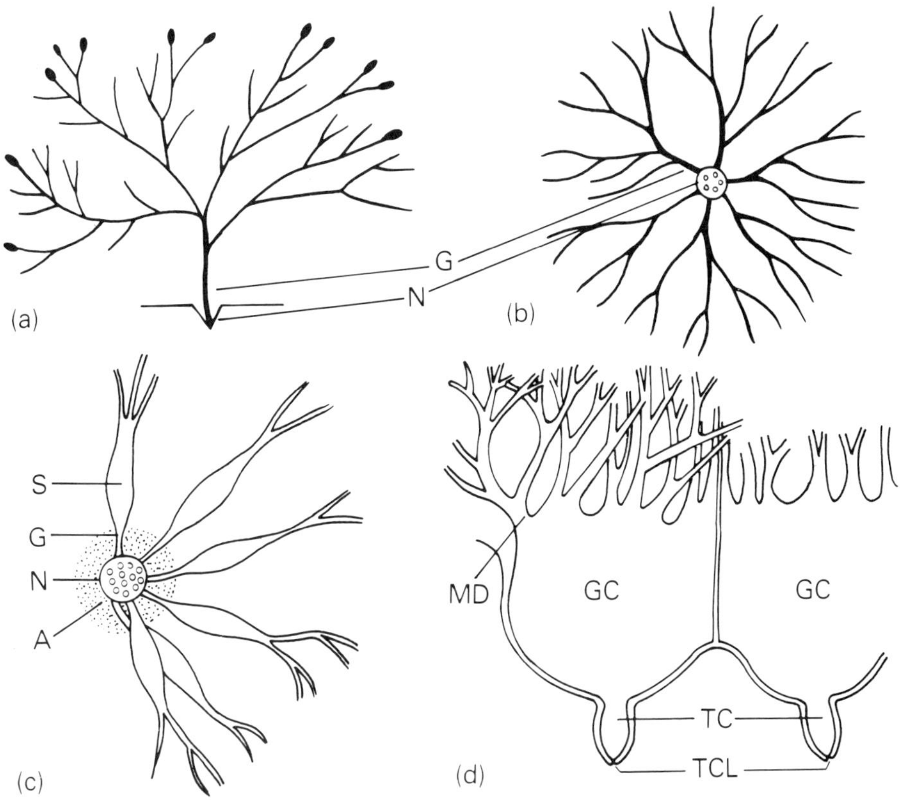

Figure 2.4. Arrangements of mammary duct systems. (a) Rat: ducts unite to form a single galactophore (G), opening at the tip of a nipple (N). (b) Rabbit: several galactophores open at nipple. (c) Woman: each galactophore is dilated at base of nipple to form a sinus (S); the nipple is surrounded by a pigmented areola (A). (d) Cow: main ducts (MD) open into a large gland cistern (GC), which communicates via a narrow aperture with the teat cistern (TC); milk is removed via the teat canal (TCL). The udder of cows comprises four separate glands, of which two are shown here.
(Source: Cowie, 1972.)

2.2.6. Nerves

The nerve supply to mammary glands, while again being largely determined by gland location, can, in general terms, be classed as of two types: (i) *somatic sensory* (afferent) *nerves*, conveying impulses from tactile receptors in the skin to the central nervous system (CNS); and (ii) *sympathetic motor* (efferent) *nerves*, which innervate blood vessels and contractile muscles of the teat. Thus, in ruminants innervation is chiefly by the external spermatic and perineal nerves, which carry sensory fibres to the CNS and also vasomotor fibres to the gland from the lumbar and sacral sympathetic chains (see Figure 2.6).

A third type of innervation may also exist. Electrophysiological and histochemical studies on rabbits suggest that there is a sparse innervation of the large ducts in the glands, which might be involved both in detection of the degree of

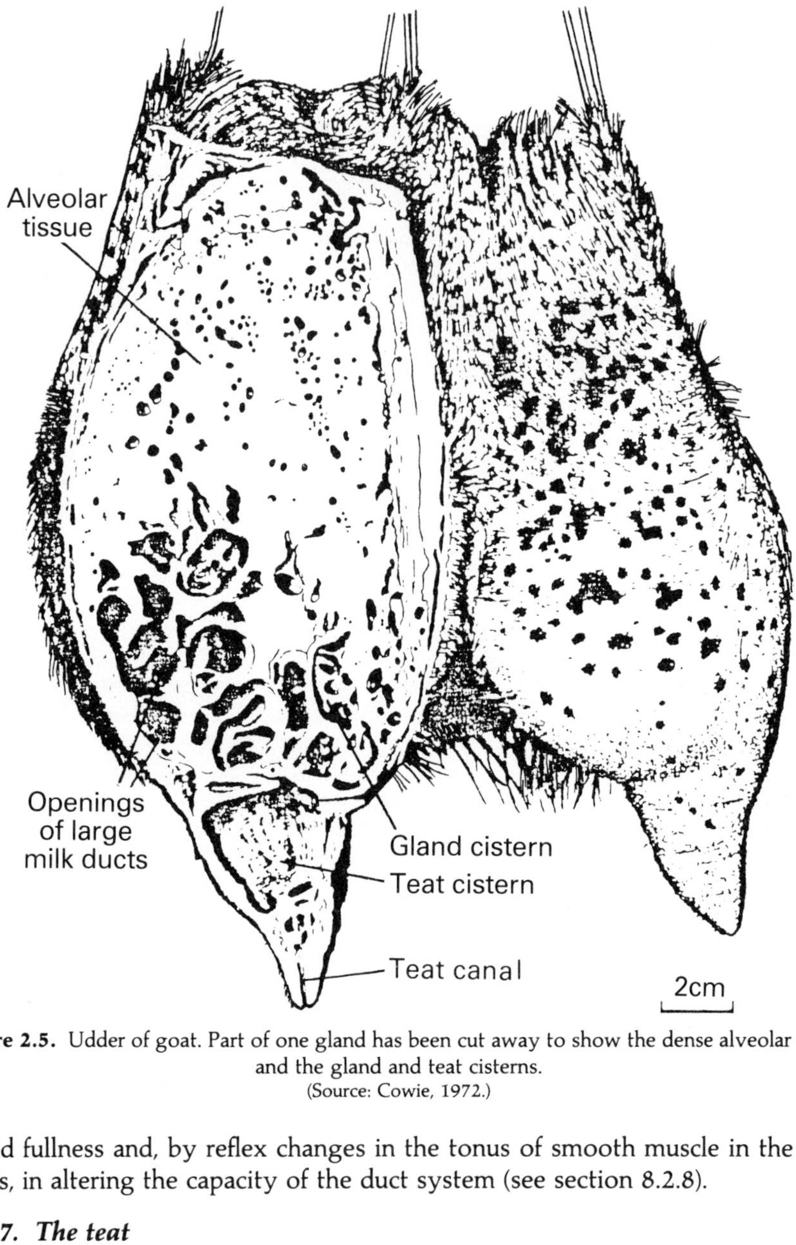

Figure 2.5. Udder of goat. Part of one gland has been cut away to show the dense alveolar tissue and the gland and teat cisterns.
(Source: Cowie, 1972.)

gland fullness and, by reflex changes in the tonus of smooth muscle in the duct walls, in altering the capacity of the duct system (see section 8.2.8).

2.2.7. The teat

In lactating ruminants the skin accounts for 10–20% of the total weight of an emptied gland. Its most obvious feature in all species except monotremes is a teat or nipple, which shows several histological characteristics related to its role of providing an efficient route for transfer of milk to the suckling young.

As briefly discussed (see section 1.5), milk removal depends on a neurohumoral reflex initiated by stimulation of tactile receptors in the teat. The teat is, by

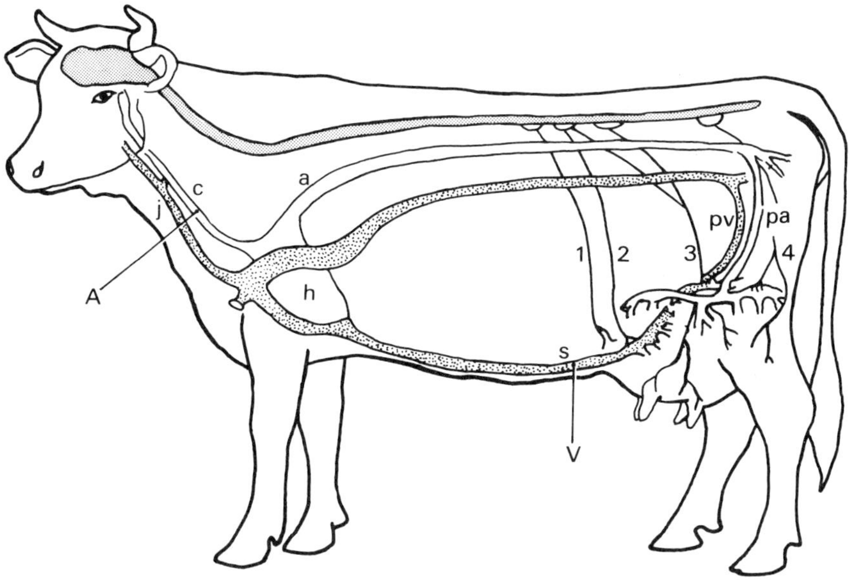

Figure 2.6. The blood vessel and nerve supply to the mammary glands of a cow. *Circulatory system* (arteries, white; veins, stippled): h, heart; a, abdominal aorta; pa, external pudic artery; pv, external pudic vein; s, subcutaneous abdominal vein; c, carotid artery; j, jugular vein. *Nerves*: 1, first lumbar nerve; 2, second lumbar nerve; 3, external spermatic nerve; 4, perineal nerve. A and V show blood sampling points for AV difference determinations (see Section 4.5).

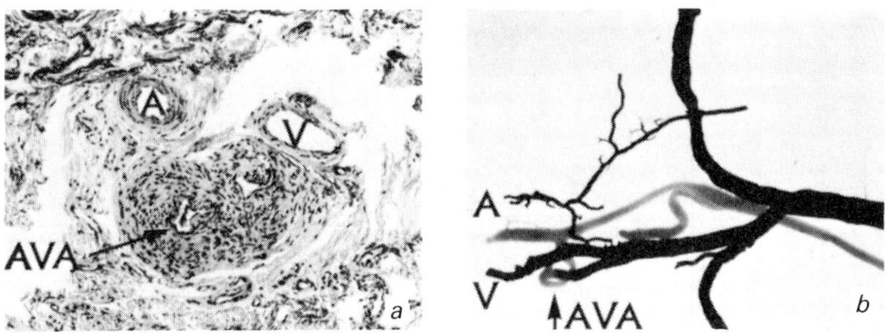

Figure 2.7. Arteriovenous anastomoses (AVA) in teat of a goat's mammary gland. (a) Transverse sections of small artery (A), vein (V) and AVA. (b) Latex preparation: red latex (grey) was injected in arteries and blue latex (black) into veins. (Scale, 100 μm).
(Reproduced from Linzell, by permission of Butterworths Medical Books.)

common observation, one of the most sensitive areas of the body surface and this is borne out by its profuse innervation, the receptors consisting of bundles of unmyelinated nerve endings situated deep in the dermis (Figure 2.8). By contrast, the rest of the mammary skin is sparsely innervated by fibres which are confined to the epidermis or superficial dermis.

During suckling the teat is subjected to marked fluctuations in pressure, such that it might be anticipated that changes in blood flow would be induced.

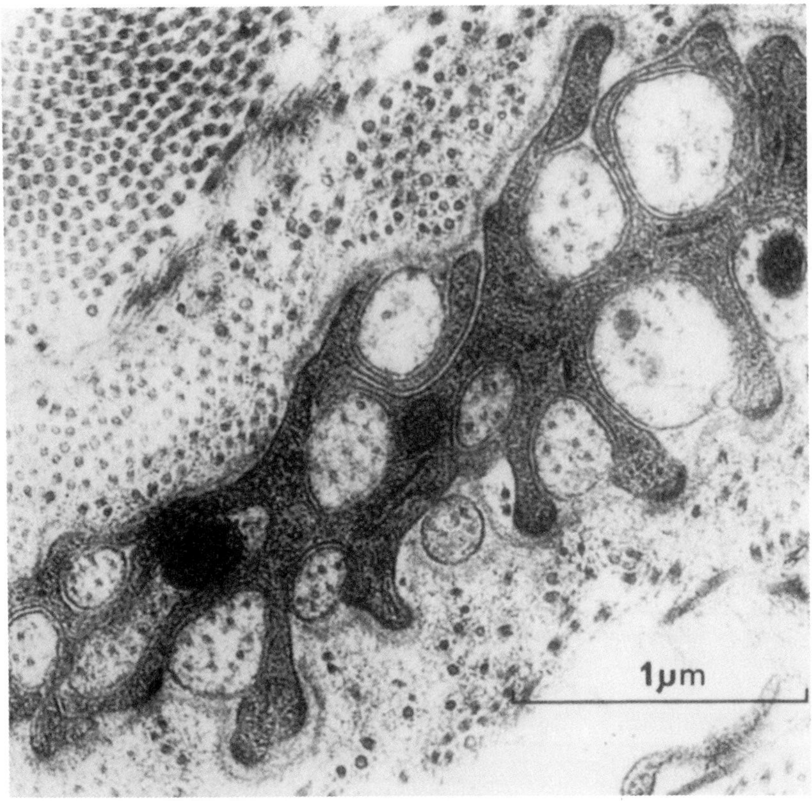

Figure 2.8. Electron micrograph of the tip of a rabbit's teat, showing a Schwann cell invaginated by unmyelinated nerve fibres lying in connective tissue.
(Reproduced from Cross and Findlay, by permission of the University of Pennsylvania Press and A. R. L. Findlay.)

However, teat veins resemble arteries in having thick muscular walls (Figure 2.9), which resist the effects of pressure changes. The teat also has a high concentration of arteriovenous shunts.

In most mammals, the teat becomes erect during suckling: this would seem to be important in enabling the young to grasp and retain it in the mouth. Teat erection is probably a consequence of smooth muscle contraction and engorgement of teat blood vessels, both of which occur in response to sympathetic nerve stimulation.

The red colour of teats in some species may be important to the newborn: some zoologists claim that redness, easily distinguishable from the pastoral greens, browns and blues of the natural environment, is a widely employed device facilitating recognition of 'significant features'. (Why else do we paint fire-engines red?)

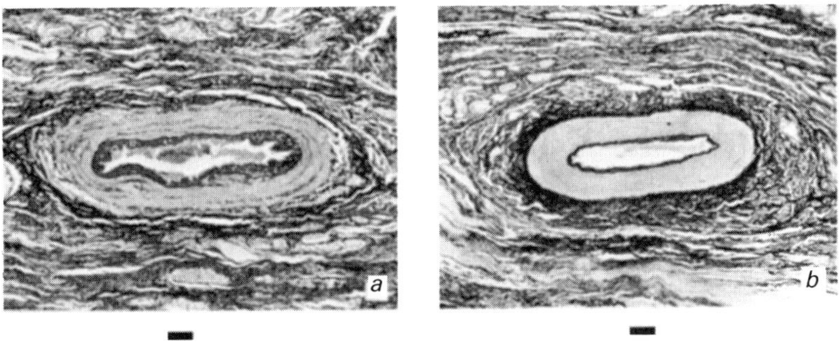

Figure 2.9. Blood vessels in the teat of a goat mammary gland. (a) vein; (b) artery. Both have well-developed muscular media and an elastic lamina, but the intima is much thicker and often irregular in veins. (Scale, 100 μm).
(Reproduced from Linzell, with permission of Butterworths Medical Books.)

2.3. Ultrastructure

Having surveyed the general features of gross morphology, we can now profitably consider the ultrastructure of the cells involved in synthesizing milk and making it available to the young: the secretory and myoepithelial cells.

2.3.1. Secretory cells

The mammary secretory epithelium is probably unique in the range of different chemical materials which it elaborates; all the fats, proteins and carbohydrates are secreted by a single type of cell. In the nineteenth century, with only the light microscope at their disposal, cytologists engaged in heated debates about cell structure and the mode of milk secretion, but with the introduction of electron microscopes (EM) in the last 30 years or so there has been a much greater measure of agreement as to cell ultrastructure and function.

The secretory cells, lining the alveolar lumen in a single layer, are of roughly cuboidal shape and markedly polarized in structure (Figure 2.10). The nucleus is usually situated at the base of the cell (towards the 'blood' side) and above it is a well-developed *Golgi apparatus*, consisting of layers of flattened vesicles. Most of the cell volume is occupied by *rough endoplasmic reticulum* (RER) and there is a large number of *mitochondria*. Other prominent features are large fat droplets and vesicles containing, what are termed from their EM appearance, 'granules' of protein. *Lysosomes*, containing digestive enzymes, are also present in the cytoplasm.

The different bounding membranes of the cell exhibit distinctive features; for example, the basal membrane shows numerous infoldings and interdigitating processes by which the surface area for substrate uptake from the ISF is increased, while the apical surface possesses microvilli which may serve the same purpose with respect to material secreted into the alveolar lumen.

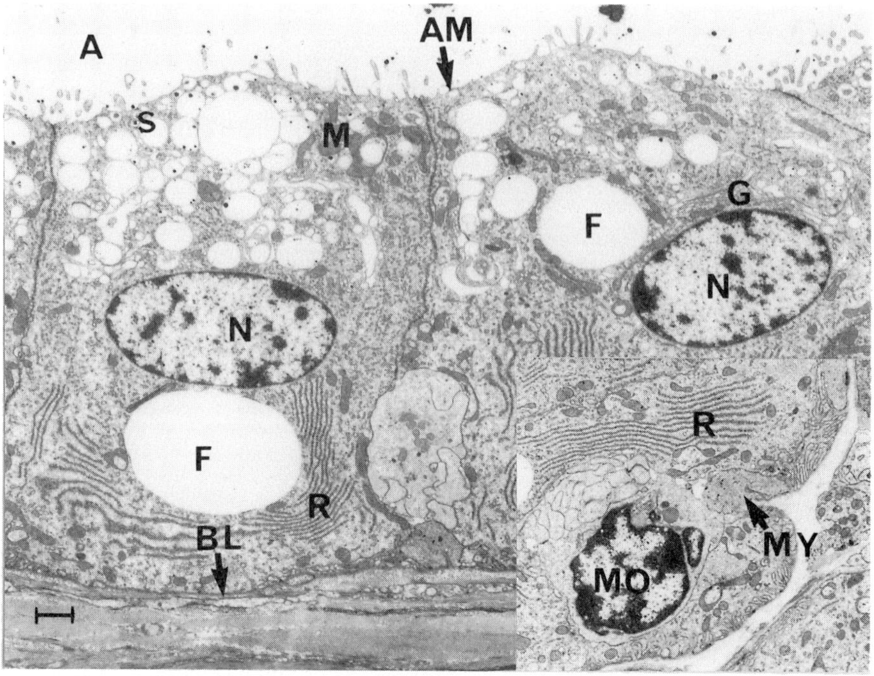

Figure 2.10. Electron micrograph of two secretory cells in bovine mammary tissue. (Scale bar, 1.0 μm). A, alveolar lumen; AM, apical membrane; BL, basal lamina; F, fat droplets; G, Golgi apparatus; S, secretory vesicle containing protein granules; N, nucleus; R, rough endoplasmic reticulum; M, mitochondria. *Inset* MO, monocyte; MY, myoepithelial cell.
(Courtesy Dr F. B. P. Wooding.)

2.3.2. Membrane junctions

The maintenance of the differential compositions of milk and ECF (that is, *extracellular fluid,* which comprises blood plasma and ISF) is ensured by the close adhesion of neighbouring cells, a condition achieved by the presence of numerous junctional complexes which 'cement' the cells together. The two classes of membrane junction are *tight junctions* (*zonula occludens*) and *gap junctions* (*macula communicans*).

2.3.2.1. Tight junctions. These (Figure 2.11) form belts encircling the apical poles of the cells so that each is sealed to its neighbours to form a continuous cellular mosaic; this arrangement seems to be an inherent property of the cells because even a suspension of disaggregated cells spontaneously takes up this configuration if placed on a collagen gel. The belt appears, on EM examination, to consist of a network of strands linking the internal plane of each membrane with that of its neighbour, thus effecting fusion of the membranes over a discrete area. Tight junctions are not, however, invariably 'tight'. Prior to parturition, and at certain

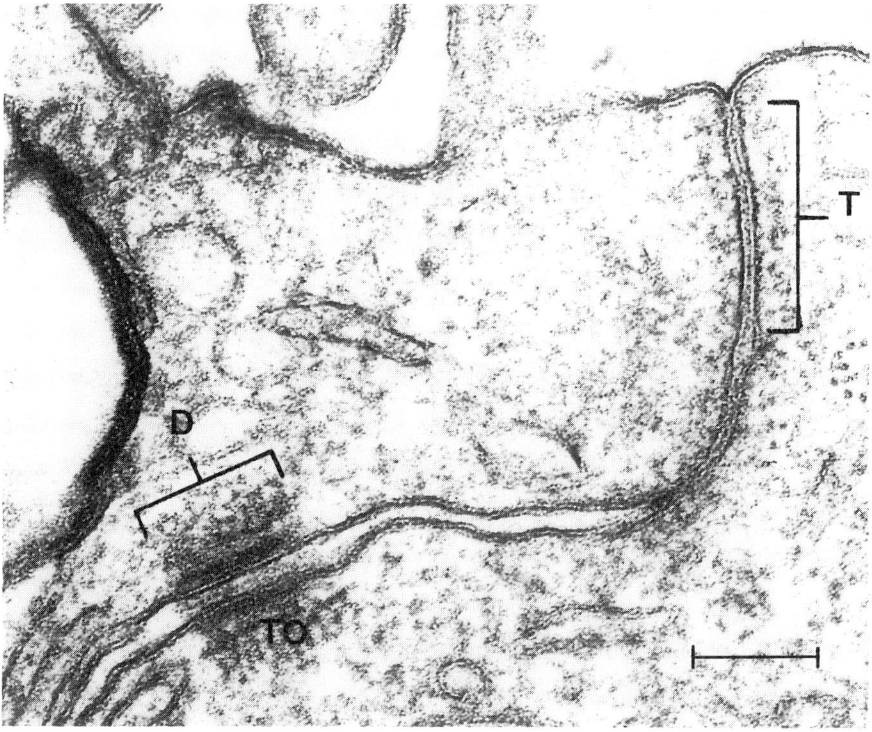

Figure 2.11. Electron micrograph of apical regions of two secretory cells in ovine mammary tissue. (Scale bar, 0.1 μm). T, tight junction; D, desmosome; TO, tonofilaments. (Courtesy Dr F. B. P. Wooding.)

times during lactation, the density of strands is reduced and the junctions become 'leaky', allowing partial equilibration of milk and ECF.

2.3.2.2. Gap junctions These are composed of 7–8 nm particles (*connexons*) which. are arranged in approximately hexagonal fashion and link two apposed cell membranes. They are present between adjacent secretory cells, between myoepithelial cells and between secretory and myoepithelial cells (see Figure 6.1, page 95). In addition to preventing leakage of material through intercellular spaces, they also appear to be a route by which ions, small molecules and electrical currents may be spread between adjacent cells, a factor of presumed significance in relation to the synchronous behaviour of the cells constituting a single alveolus.

2.3.3. *Desmosomes*

These are a second, distinctive, type of cell junction. They are recognizable by the presence of dense material under and between two cell membranes, which are separated by 10–30 nm; the membranes are thickened, and dense filaments radiate out into the cytoplasm. The function of desmosomes appears to be solely that of maintaining cell adhesion. In mammary tissue, desmosomes are present as *spots* (less than 2 μm in diameter, also called *macula adhaerentes*), and *belts* (*intermediate junctions*, also called *zonulae adhaerentes*), the latter encircling the cells just below the tight junction (see Figure 2.11). They are best developed in the mammary ducts, where in the form of spots they connect ductal and myoepithelial cells in both like and unlike pairs. Hemidesmosomes, which involve only one membrane and look like half a desmosome, seem primarily to be sites of attachment of myoepithelial and epithelial cells to the basal lamina (see section 2.3.7) and fibrous stroma.

2.3.4. The cytoskeleton

The shape and polarization of the epithelial cells implies the existence of an extensive system of structural support. This is provided by the cytoskeleton, an intricate network of filaments which is firmly attached to the inner surface of the plasma membrane and envelopes, and is continuous with, the various cytoplasmic organelles. As will be discussed more fully (see section 6.8), the cytoskeleton is in a continuous state of flux and many of the dynamic properties of cells, such as motility, intracellular translocation of molecules and cell-division, are dependent on its structure and physicochemical properties.

The cytoskeletal filaments, largely protein in composition, fall into several groups on the basis of their average diameters: *actin* (6 nm); *tonofilaments*, composed of *keratin* (10 nm) and *microtubules*, composed of *tubulin* (24 nm). To date, it has proved difficult to visualize the precise interrelationships between the different types of filament, so that no clear understanding of their interaction has emerged. However, a topographical description has been possible when the filaments aggregate in dense bundles. For example, microfilaments of actin form a core to each apical microvillus and this continues as a web of actin and keratin filaments beneath the apical surface; actin originating from this web is attached to the membranes at tight junctions, while keratin is associated with belt-desmosomes and spot-desmosomes (Figure 2.11). Microtubules are common in the Golgi region, where their contractile properties seem to be important in the intracellular transport of protein-containing vesicles to the apical cell membrane (see section 6.2).

2.3.5. Ductal cells

There is no abrupt division between secretory and ductal cells, since cells lining intralobular ducts seem capable of secreting milk. Nevertheless, most ductal cells are non-secretory and present a distinctive appearance by EM.

The cells are roughly cuboidal and have a relatively undifferentiated cytoplasm. Thus, RER is sparse (although there are many free ribosomes) and the Golgi apparatus is small. There are some reports of a double layer of epithelial cells plus a myoepithelial cell layer lining ducts, but other reports suggest that only one epithelial cell layer is present.

Since milk does not appear to undergo significant changes in its chemical composition when stored in the ducts, despite the marked degree of stretching to which the latter are sometimes exposed, it might be anticipated that membrane complexes play as important a role as they do in alveoli. This is substantiated by the presence of well-developed tight junctions and belt-desmosomes and spot-desmosomes (see section 2.3.3).

2.3.6. Myoepithelial cells

Myoepithelial cells were first described at the end of the nineteenth century but their significance was not generally recognized and it was only in the 1940s that they were rediscovered and their importance in milk ejection (see section 1.5) fully appreciated. Ultrastructurally, they resemble smooth muscle cells, the cytoplasm being replete with filamentous tracts and plasmalemmal caveolae. The tracts are composed of actin filaments, a loose meshwork of intermediate filaments and scattered dense nodes. Mitochondria, RER and a few fat droplets are chiefly confined to the perinuclear region.

The arrangement of the myoepithelial cells around the alveoli is clearly shown by silver staining (Figure 2.12), and is also evident in the scanning electron

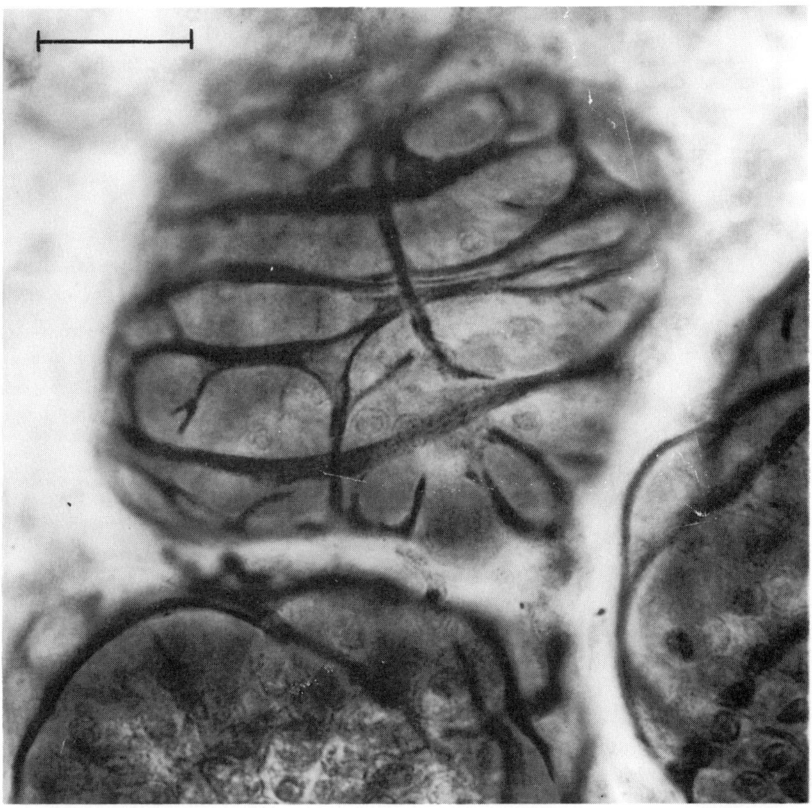

Figure 2.12. Myoepithelial cells surrounding alveoli in goat mammary tissue (Scale bar, 25 μm). (Reproduced from Richardson, 1949, by permission of The Royal Society and K. C. Richardson.)

micrograph (see Figure 2.2) following removal of superficial connective tissue. A cross-section of one such cell is also shown in Figure 2.10. There appear to be about five such cells per alveolus in the rat mammary gland, each giving rise to extended processes so that the myoepithelium of each alveolus presents a basketwork appearance. Since milk is expelled from the alveoli by contraction of these cells, it would be anticipated that efficient expulsion depends on their coordinated contraction. This does not appear, however, to be achieved by cellular continuity, as in heart muscle, for example, where the cells form a syncytium. A more likely explanation is that gap junctions, which have been identified in this tissue, allow a spread of regulatory ions and molecules between contiguous cells, thus ensuring coordinated contraction of all the cells in an individual alveolus.

2.3.7. *The basal lamina*

Surrounding each alveolus in the form of a smooth blanket, is the basal lamina (basement membrane), a layer of material (chiefly type IV collagen, glycoprotein and glycosaminoglycans) secreted by the epithelial cells at their interface with the connective tissue matrix (see Figure 2.10). By EM it appears to comprise a fibrous and granular layer, about 30 nm thick, and a lower-density layer between the latter and the epithelial plasmalemma. The basal lamina provides a barrier between the epithelial and stromal compartments, which is not crossed by cells other than migrating leucocytes.

2.3.8. *Leucocytes*

Electron microscopic examination of mammary tissue often reveals the presence of lymphocytes or monocytes in alveoli, wedged between two adjacent epithelial cells (see Figure 2.10). Such cells, having migrated to the mammary gland (see section 2.3.7.), play a role in local production of antibodies (see section 3.3.3).

2.3.9. *Adipose tissue cells*

These cells are present in the interlobular connective tissue. In non-lactating animals, adipose tissue accounts for a large proportion of the mammary volume, but, with the lobuloalveolar development of pregnancy, fat is mobilized so that adipocytes in lactating tissue contain little fat.

Further reading

Cowie, A.T. and Tindal, J.S. (1971). Physiology of the mammary gland, in *The Physiology of Lactation*, pp. 3–34. (London: Edward Arnold.)

Hollman, K.H. (1974). Cytology and fine structure of the mammary gland, in *Lactation* Vol. 1, pp. 3–96, eds, Larson, B.L. and Smith, V.R. (New York and London: Academic Press.)

Gould, S.F. (1983). Anatomy of the breast, in *Lactation: Physiology, Nutrition and Breast Feeding*, pp. 23–47. eds Neville, M.C. and Neifert, M.R. (New York and London: Plenum.)

Linzell, J.L. (1971). Mammary blood vessels, lymphatics and nerves, in *Lactation*, pp. 41–50. ed. Falconer, I.R. (London: Butterworths.)

Linzell, J. L. (1974). Mammary blood flow and methods of identifying and measuring precursors of milk, in *Lactation*. Vol. 1, pp. 143–225, eds Larson, B. L. and Smith, V. R. (New York and London: Academic Press.)

Long, C. A. (1969). The origin and evolution of mammary glands, *Bioscience* **19 (6)**, 519–23.

Nagato, T., Yoshida, H., Yoshida, A. and Uehara, Y. (1980) A scanning electron microscope study of myoepithelial cells in exocrine glands. *Cell Tissue Res.* **209**, 1–10.

Pitelka, D. R. (1978). Cell contacts in the mammary gland, in *Lactation* Vol. 4, pp. 41–66. ed. Larson, B. L. (New York and London: Academic Press.)

Pitelka, D. R. and Hamamoto, S. T. (1983) Ultrastructure of mammary secretory cells, in *Biochemistry of Lactation*, pp. 29–70. ed. Mepham, T. B. (Amsterdam and New York: Elsevier).

Schmidt, G. H. (1971) Mammary gland anatomy, in *Biology of Lactation*, pp. 6–35. (San Francisco: W. H. Freeman).

Wooding, F. B. P. (1977) Comparative mammary fine structure *Symp. Zool. Soc. Lond.* **41**, 1–41.

3 The Composition of Milks

3.1. Introduction

At the time of its secretion, milk comprises two liquid phases, between which are partitioned a large number of substances. In the aqueous phase, lactose, mineral ions and water-soluble vitamins are present in simple solution, while proteins and other ions are present in colloidal suspension. The lipid phase contains fat-soluble vitamins and compounds such as sterols and carotenoids. There is a basic qualitative similarity in the composition of milks of different species in that all contain certain distinctive types of fat, protein and carbohydrate; but there is considerable interspecies variation in the precise nature of these substances and their relative concentrations.

Since milk is the sole food for the *neonate* (newborn) in the immediately postpartum period (although this period may be as short as a few days, for example in guinea-pigs, or as long as months, for example in monotremes), it might be imagined that it would contain all the nutrients necessary for growth and activity, but this is not necessarily the case. Thus, most milks contain little iron, and in some animals haemoglobin is formed from iron reserves in the liver which were accumulated *in utero*. In piglets, iron reserves are inadequate, and supply depends on the young rooting in the soil almost immediately after birth. In modern production systems, where pigs are reared under artificial conditions, anaemia is prevented by administration of iron by injection. Again, under natural conditions, vitamin D is formed in the skin through the action of sunlight, and milk itself contains amounts which are inadequate for bone and tooth formation. While such potential deficiencies need to be recognized when animals are bred under artificial conditions, it remains true that milk is an almost complete food, and, certainly, the most complete *single* food.

3.2. Milk as a food

The distinctive feature of milk as a food lies in its balanced content of essential nutrients. This has been appreciated for thousands of years and led to the early domestication of docile, herbivorous ruminants, like cows, goats, camels and reindeer—animals which, because they are able to digest the cellulose of plant cell walls, do not compete with humans or other monogastric animals for food (see sections 5.2.1 and 11.3).

The major milk components of a number of species are shown in Table 3.1.

Table 3.1. Average concentrations of some constituents in the milk of a number of species. (*Source: Davies, Holt and Christie, 1983.*)

Species	*Fat (g/l)*	*Casein (g/l)*	*Milk serum protein (g/l)*	*Lactose (mM)*	*Calcium (mM)*
Cow	37	28	6	133	30
Goat	45	25	4	114	22
Sheep	74	46	9	133	58
Pig	68	28	20	153	104
Horse	19	13	12	172	17
Man	38	4	6	192	7
Rat	103	64	20	90	80
Guinea-pig	39	66	15	83	41
Rabbit	183	104	32	60	214

As we shall see, the particular chemical compositions of milks of different species are related to characteristics such as stage of maturity at birth and environmental conditions, so that it is unlikely that milk of any one species would be ideally suited to the young of another species. But for weaned animals, which need be far less fastidious, consumption of milk from other species, in appropriate moderation, would usually appear to have few drawbacks and many advantages. For example, 0.5 l of cow's milk supplies about 25% of the energy, 30% of the vitamin A and thiamin, 40% of the protein and 70% of the calcium and riboflavin required by a 5-year-old child. Milk is also valuable for the elderly, who, prone to *osteoporosis* (brittle bones, due to mineral depletion), benefit from its high calcium and phosphorus content. The importance of milk and its various products (cheese, butter, yoghurt, etc.) in the diet of people in western countries is indicated by the fact that the average consumption of milk (in all its forms) per person in the EEC is about 1 litre per day. Members of the nomadic M'bororo tribe in West Africa are reputed to live exclusively on milk for periods of several months.

Like many 'health foods' today, milk has been the subject of fads. B. Macfadden in *The Miracle of Milk* (1926) advocated the, quite intemperate, drinking of milk to cure everything from asthma to piles, and from rheumatism to impotence. He recommended consumption of up to 12 pints (6.8 l) daily, i.e. 'a quart (1.14 l) for each foot (30 cm) of height'! Nowadays we are more circumspect; we know all too well that excessive saturated fat is 'bad for the arteries'. Another significant restriction on the use of milk for a large proportion of the human population is *lactose intolerance*. The majority of people not of European ethnic origin (see section 11.3.1) are lactose malabsorbers as a result of the decline in intestinal *lactase* activity which normally occurs in early childhood. The condition is, however, rarely absolute, and milk products such as yoghurt and cheese can often be tolerated (see section 11.3.3 J).

In the following sections we shall consider the chemical composition of milk in some detail. This is important in establishing both the basis of the nutritional properties of milk and the manner in which the mammary cells are engaged in its biosynthesis. Most of the figures to be quoted refer to cow's (*Bos taurus*) or human milk. These species have been chosen because of their obvious importance in human nutrition, but it should be appreciated that such milks are in no sense representative of milks in general, of which only about 5% have been subjected to any sort of analysis. Indeed, with its low protein and high lactose concentrations, human milk is highly atypical. It must also be remembered that figures quoted cannot be regarded as definitive for the species because milk composition varies in response to the condition of the animal and, in particular, to the stage of lactation (see section 3.10).

3.3. Proteins

In broad terms we can classify milk components into one of four groups (organ here refers to the mammary gland):

(a) Organ-specific and species-specific
(b) Organ-specific but not species-specific
(c) Not organ-specific but species-specific
(d) Neither organ-specific nor species-specific.

For milk protein, another simple classification is useful: *caseins*, on the one hand, and *milk serum* (*whey*) *proteins*, on the other. Caseins are organ-specific phosphoproteins, with molecular weights usually in excess of 20 000, which are precipitated at acid pH (pH 4.6 for cow's milk) or by the action of gastric enzyme *chymosin* (also called *rennin*). Milk serum proteins are those which remain in solution at the pH of casein precipitation.

3.3.1. Caseins

In those caseins subjected to detailed study (such as cow's milk), virtually all the phosphate is combined with seryl residues of the polypeptide chain as o-phosphomonoesters. A large number of prolyl residues are widely distributed along the polypeptide chain, producing an 'open' tertiary structure which facilitates digestion by proteolytic enzymes.

Using the technique of gel electrophoresis, it has been shown that casein consists of a mixture of different proteins which, in cow's milk, are classifiable into four types: α_s, β-, γ- and κ- caseins. Even these types of casein are not homogeneous, polymorphism resulting from (a) genetic variation and (b) a variable degree of modification of the molecules, for example, by phosphorylation. When it is realized that α_s-casein may be further characterized as α_{s1} and α_{s2} and that four genetic variants of α_{s1} have been identified, some idea may be gained of the complexity of the problem of describing milk proteins as a whole.

Table 3.2. Contents and composition of protein in cow's milk. *Values are means of 29 samples taken from five dairies. SD, standard deviation; IPL, immunoglobulins, proteose-peptone components and lactoferrin. (Source: Davies, Holt and Christie, 1983.)*

	Amount (g/l) in skim-milk		*Per cent of total protein*		*Per cent of total casein*		*Per cent of total milk serum protein*	
	Mean	*SD*	*Mean*	*SD*	*Mean*	*SD*	*Mean*	*SD*
Total protein	32.71	1.80	–	–	–	–	–	–
Total casein	26.92	1.54	82.2	0.6	–	–	–	–
Total milk serum protein	5.79	0.32	17.8	0.6	–	–	–	–
α_{s1}-Casein	10.25	0.57	31.3	0.5	38.1	0.5	–	–
β-Casein	9.60	0.50	29.3	0.6	35.7	0.8	–	–
κ-Casein	3.45	0.32	10.5	0.5	12.8	0.6	–	–
α_{s2}-Casein	2.74	0.21	8.4	0.5	10.2	0.8	–	–
γ-Casein	0.88	0.15	2.7	0.4	3.2	0.4	–	–
β-Lactoglobulin	3.14	0.19	9.6	0.4	–	–	54.2	1.3
α-Lactalbumin	1.23	0.09	3.8	0.3	–	–	21.2	1.4
IPL	0.97	0.10	3.0	0.2	–	–	16.8	1.2
Bovine serum albumin	0.45	0.04	1.4	0.1	–	–	7.8	0.6

A list of the caseins of cow's milk is given in Table 3.2 (in other species the relative concentrations may be very different). Since caseins are relatively hydrophobic proteins, they exist in milk at ambient temperature in the form of colloidal particles, termed *micelles*. These are roughly spherical structures, 100–300 nm in diameter, in which α_{s1}-, α_{s2}-, β- and κ-caseins are present in roughly the ratio 3:1:3:1. But about 10% of the material in the micelles consists of calcium and inorganic phosphate, together with smaller amounts of citrate and magnesium. Most of the calcium is combined directly with inorganic phosphate within the micelle but about 25% is bound to the polypeptide chain and to citrate. It appears that micelles are composed of submicelles of calcium caseinate (Figure 3.1a) bound together by calcium phosphate (Figure 3.1b), and κ-casein has an important role in maintaining micellar structure. When κ-casein is hydrolysed by the action of chymosin its stabilizing property is lost and casein precipitates in the form of its characteristic curd.

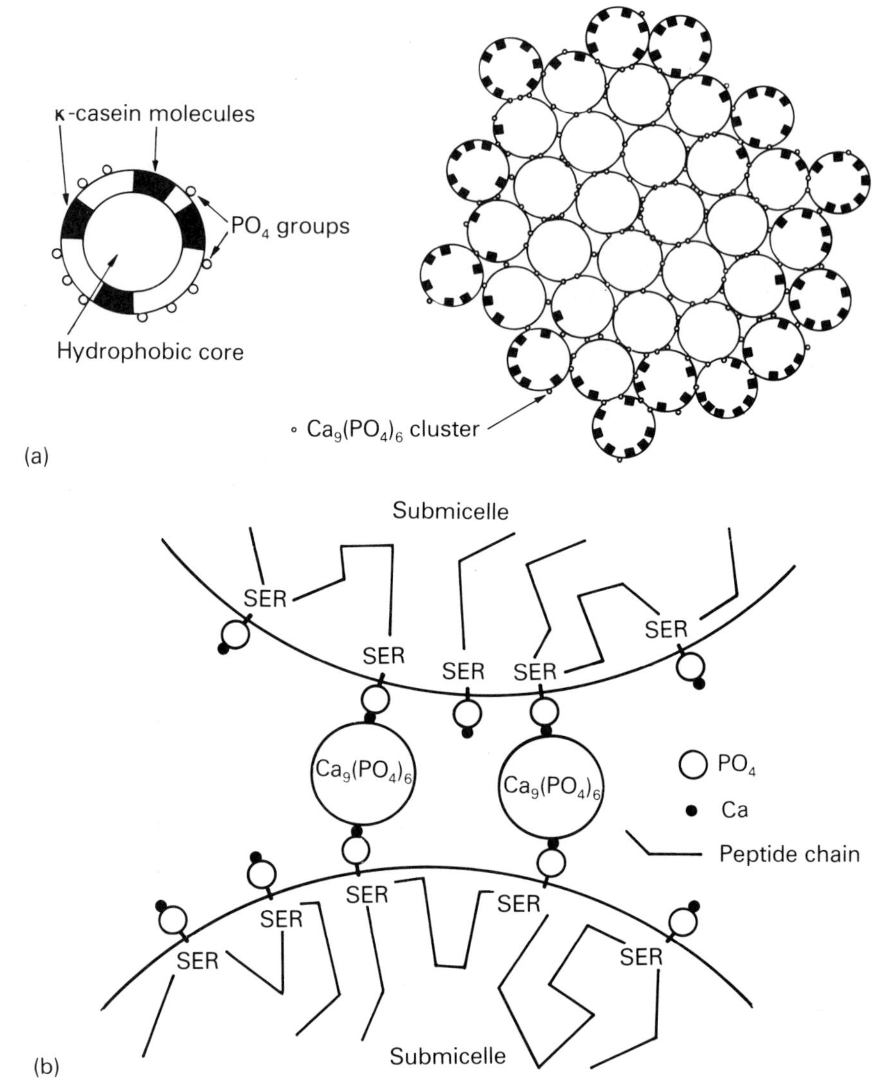

Figure 3.1. (*a*) Schematic representations of a casein micelle (right) and one of its constituent submicelles (left). (Submicelles are only approximately spherical). (*b*) Schematic representation of the binding of two submicelles via $Ca_9\,(PO_4)_6$ clusters.
(Source: Schmidt, 1982.)

3.3.2. Milk serum proteins

This group includes several kinds of protein (see Table 3.2), some of which (e.g. β-lactoglobulin and α-lactalbumin) are organ-specific, while others (such as immunoglobulins and serum albumin) are identical with blood proteins (categories (c) and (d) in section 3.3).

Beta-lactoglobulin is the principal serum protein in the milk of cows, goats and sheep but it does not occur in many other milks, including human. On digestion it yields high concentrations of essential amino acids, but it may also play another role through its ability to bind large amounts of vitamins. Alpha-lactalbumin is present in virtually all milks, a fact attributable to its participation in the catalysis of a crucial step in the pathway of lactose synthesis (see section 5.4.2).

3.3.3. *Immunoglobulins*

The nutritional properties of milk have tended to overshadow the importance of the mammary secretion of antibodies. Yet in many species, the latter, in the form of immunoglobulins, are a vital element in the protection of the newborn from the influence of infections and toxic agents. Immunology is a complex subject and there is room here only for the barest outline of the essential features as they apply to lactation.

The immunoglobulins (Igs) are a heterologous group of glycoproteins synthesized by cells of the lymphocyte plasma series. The basic unit of all Ig molecules consists of two *heavy* and two *light* polypeptide chains, which are linked by disulphide bridges in the shape of a 'Y' (Figure 3.2): different 'arms' of the molecule have specific properties, for example, acting as an antibody-binding site or as a site for complement fixation. The most prevalent Ig in blood serum is IgG, which exists as a monomer in two forms (IgG_1 and IgG_2), whereas IgA is a dimer and IgM a pentamer (see Figure 3.2). In mammary secretions IgA combines with another glycoprotein (*secretory component*) to form a complex *secretory IgA* (SIgA).

The relative concentrations of the different Igs is highly species-variable (Table

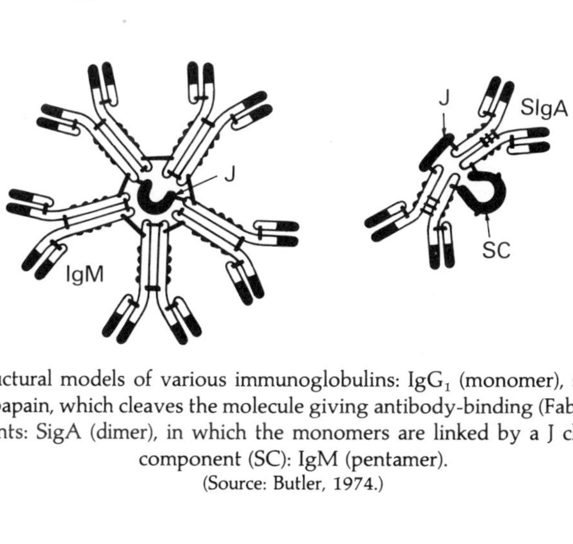

Figure 3.2. Structural models of various immunoglobulins: IgG_1 (monomer), showing the effect of treating with papain, which cleaves the molecule giving antibody-binding (Fab) and complement-fixing Fc fragments: SigA (dimer), in which the monomers are linked by a J chain and secretory component (SC): IgM (pentamer).
(Source: Butler, 1974.)

Table 3.3. Immunoglobulins in colostrum and milk of three species. *FSC, free secretory component, that is, unbound to IgA. (Source: Jenness, 1982.)*

	Concentration (g/litre)					
	Human		Pig		Cow	
Ig	*Colostrum*	*Milk*	*Colostrum*	*Milk*	*Colostrum*	*Milk*
IgA	17.4	1.0	10.7	7.7	3.9	0.14
IgG	0.4	0.04	58.7	3.0		
IgG_1					47.6	0.6
IgG_2					2.9	0.02
IgM	1.6	0.1	3.2	0.3	4.2	0.05
FSC	2.1	+			0.2	0.05

3.3) and it is clear that concentrations are much higher in colostrum than in mature milk. In some species, Igs are transferred to the embryo *in utero* via the placental membranes; in others transfer is solely via colostrum and milk, while in yet others both routes operate. As a generalization, in those species in which colostral transfer is dominant, the principal Ig is IgG, while in those where placental transfer occurs colostral concentrations of IgA are higher.

The IgG in cow's colostrum and milk is derived directly from blood, and since IgG_1 is present at much higher concentration than IgG_2 (whereas in blood their concentrations are similar), it is believed that IgG_1 is selectively transported across the mammary epithelium (see section 6.9). By contrast, IgA is synthesized by lymphocytes located adjacent to the mammary epithelium (see section 2.3.8; Figure. 2.10 (inset)). This local production of IgA is indicated by the observation that infusion of antigenic material into one of the two glands of ewes (via the teat canal) leads to much higher IgA production in the infused, than in the non-infused, gland. Thus, two sorts of immunity are involved—*passive*, in which the transfer of Ig confers on the newborn resistance to infections to which the mother is resistant, and *active*, whereby antigens introduced into the mammary gland from the mouth of the suckling young elicit a local production of antibody. For the passive immunity to be effective it is necessary for the intestinal mucosa of the newborn to be permeable to Ig molecules. There is a wide interspecies variation in the post partum period over which the gut wall is permeable to Igs, for example 24–36 h in ruminants, 16–20 days in rodents and throughout the whole period when the young remain in the pouch in marsupials (e.g. up to 200 days). On the other hand, the SIgA system is believed to act at the level of the intestinal mucosal surface by providing, what has been called, a protective 'molecular paint'.

3.3.4. *Other milk serum proteins*

Milk of many species contains a protein apparently identical to the blood serum albumin; this amounts to about 1% of total milk protein in mature cow's milk. Both *lactoferrin* and *transferrin* (which is identical to blood serum transferrin) are milk proteins which bind iron; they also have bacteriostatic properties. Several proteose-peptides are present as a result of proteolytic cleavage of milk proteins: the γ-casein of milk is formed from β-casein by this process. A large number of enzymes have been detected in milk (about 40 in cow's milk) and it is clear that their continuing activity after milk has been secreted may lead to compositional changes in the milk.

3.3.5. *Dietary significance*

For adults, cow's milk provides a palatable and highly digestible form of high quality protein. In developed countries, milk protein accounts for 20–30% of the dietary protein intake, but in India the figure is 10% or less, and elsewhere in Asia it is less than 2%.

In assessing nutritional value, the amino acid which is at lowest concentration in respect of nutritional requirements determines the *chemical score* of the protein. Vegetable proteins tend to have low scores because of their low contents of the essential amino acids, methionine, lysine and tryptophan, whereas the score for milk and egg proteins is high (Table 3.4). The fact that casein binds large amounts of calcium and phosphorus means that milk is a major source of these minerals; for example, in developed countries, over two-thirds of the dietary intake of calcium may be derived from milk.

Table 3.4. Amino acid contents of milk and other proteins. *Values are expressed as mg amino acid per g total nitrogen. (Source: Hambraeus, 1982.)*

	Human breast milk	*Cow casein*	*Cow whey protein*	*Egg*	*Beef*	*Wheat (whole grain)*	*Soybean*
Isoleucine	254	345	476	393	301	204	284
Leucine	471	607	736	551	507	417	486
Lysine	337	518	704	436	556	179	399
Methionine	78	178	151	210	169	94	79
Cystine	114	23	174	152	80	159	83
Phenylalanine	171	334	224	358	275	282	309
Tyrosine	223	371	214	260	225	187	196
Threonine	228	297	527	320	287	183	241
Tryptophan		103	147	93	70	68	80
Valine	296	430	449	428	313	276	300
Arginine	171	239	175	381	395	288	452
Histidine	114	186	144	151	213	163	158
Alanine	166	196	341	370	365	226	266
Aspartic acid	451	455	766	601	562	308	731
Glutamic acid	1 000	1 406	1 231	796	955	1 866	1 169
Glycine	98	126	126	207	304	245	261
Proline	513	738	450	260	236	621	343
Serine	228	385	374	478	252	287	320

There are, however, marked differences in the protein content of cow's and human milks, and this has important implications for the design of artificial feeds for babies. For example, the total protein content of human milk is less than one-third of that in cow's milk, while the casein:whey ratio is about 82:18 in cow's milk and 30:70 in human milk. Attempts to modify (*humanize*) cow's milk for use by babies thus face formidable problems (Figure 3.3). In fact, the problems are incapable of satisfactory solution because the specific proteins of the two milks are quite different, for example, in terms of their amino acid contents, and in the types of curd formed when casein is precipitated. It has been suggested that β-lactoglobulin (which is absent from human milk) may be responsible for the allergic reactions which some infants show to cow's milk formula feeds. Moreover, recent evidence suggests that low-birthweight infants often have particular metabolic deficiencies which would be seriously aggravated by the use of cow's milk formula.

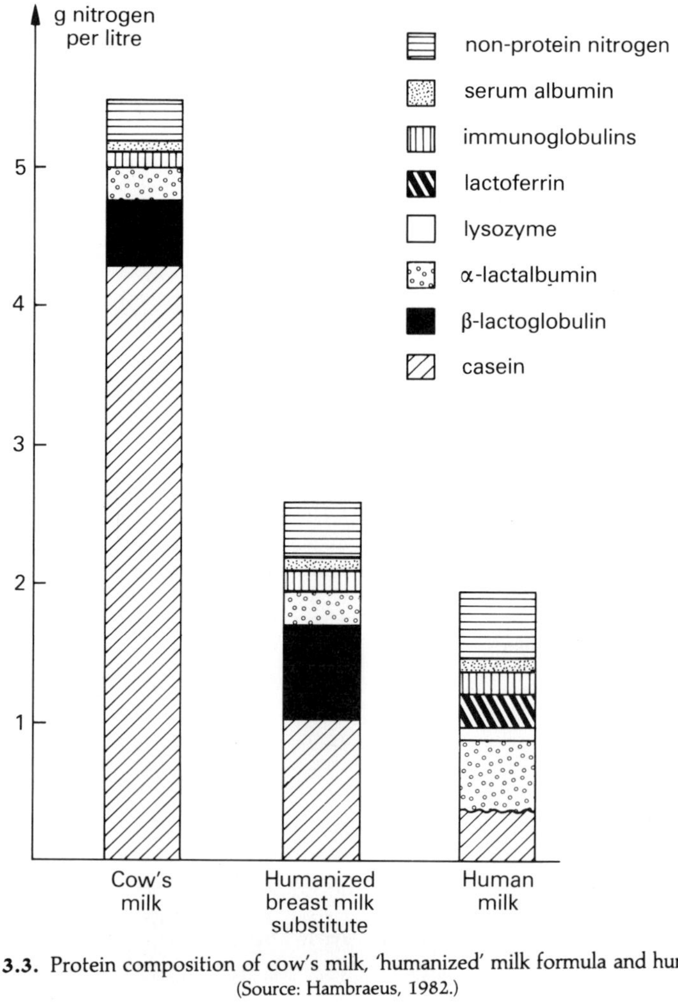

Figure 3.3. Protein composition of cow's milk, 'humanized' milk formula and human milk. (Source: Hambraeus, 1982.)

3.4. Milk fats

Fats are the most variable component of milk in both their concentration and chemical composition, whether interspecies or intraspecies differences are considered. The milks of some species contain little or no fat (for example, in black rhinoceros); while those of aquatic species (such as seals) are over 50% fat by weight. The main class of fats is the *triacylglycerols* (TG), which may account for over 98% of total lipids; the compositional variation is principally due to the wide range of fatty acids which may be esterified with glycerol in this lipid fraction. As many as 437 different fatty acids have been isolated from cow's milk, and if all these were randomly esterified among the three positions of the glycerol backbone a total of over 8.3×10^7 individual TG species would be possible! However, in quantitative terms, the most important of the fatty acids in the TG fraction comprise a list of 14 (Table 3.5), all of which are readily separable by conventional gas chromatography techniques. The remainder of the lipid fraction (approx 2% of the total in human and cow's milk) consists mainly of diacylglycerols, cholesterol, phospholipids and free fatty acids (Table 3.6).

Table 3.5. Fatty acid compositions of the TG fraction of milk of cows, human females and rats, *expressed as weight/per cent of total fatty acids. (Source: Davies, Holt and Christie, 1983.)*

Fatty acid	*Cows*	*Human females*	*Rats*
4:0	3.3		
6:0	1.6		
8:0	1.3		1.1
10:0	3.0	1.3	7.0
12:0	3.1	3.1	7.5
14:0	9.5	5.1	8.2
15:0	0.6	0.4	
16:0	26.3	20.2	22.6
16:1	2.3	5.7	1.9
17:0	0.5		0.3
18:0	14.6	5.9	6.5
18:1	29.8	46.4	26.7
18:2	2.4	13.0	16.3
18:3	0.8	1.4	0.8

Some generalizations about the TG fraction are noteworthy:

(a) Most milks contain a much higher proportion of short-chain and medium-chain length fatty acids than is present in the other tissues of the same species.

(b) Milks of ruminant species have a relatively higher concentration of short-chain fatty acids ($<$C9) than is the case for non-ruminants; they also contain much less of the polyunsaturated fatty acids (18:2 and 18:3) than non-ruminant milks.

Table 3.6. Compositions of the main lipid classes of the milk of cows, human females and rats, expressed as weight/per cent of total lipids. *(Source: Davies, Holt and Christie, 1983.)*

Lipid class	*Cows*	*Human females*	*Rats*
Triacylglycerols	97.5	98.2	87.5
Diacylglycerols	0.36	0.7	2.9
Monoacylglycerols	0.027	trace	0.4
Cholesterol esters	trace	trace	
Cholesterol	0.31	0.25	1.6
Free fatty acids	0.27	0.4	3.1
Phospholipids	0.6	0.26	0.7

(c) The particular fatty acid composition of milk TG is dependent on the nature of the diet of the lactating animal to a greater degree than is the case for other milk constitutents.

One of the practical repercussions of knowledge of the fatty acid composition of milk fats is that artificial milk powders manufactured from cow's milk are supplemented with vegetable oils to increase the linoleic acid (18:2) content, which otherwise might be inadequate for human infants (see Table 3.5).

Milk fat is secreted in the form of spherical globules (see Figure 6.8), which are the largest particles present in milk and easily visible by light microscopy: in cow's milk they range from 0.1 to 20 μm in diameter. The *milk fat globule membrane* (MFGM) surrounding the globules has been studied intensively by chemists, both because it provides important clues as to the nature of milk fat secretion (see section 6.7) and because its physicochemical properties have consequences for treatments employed in the preservation and processing of milk and milk products. Phospholipids are integral parts of the MFGM and protein constitutes a major component. Although TG adheres to the membrane during isolation procedures it is not thought to be an integral part of it.

3.4.1. Dietary significance

Dietary lipids have three broad types of function, that is, they act in *structural, metabolic* and *storage* roles. Thus, lipids serve as components of cell membranes, act as precursors of specific compounds (for example, steroid hormones) and constitute a major store of high-energy-yielding substrate: (the energy values of 1.0 g of fat, protein and carbohydrate are 37 kJ, 17 kJ and 16 kJ, respectively). In Britain, milk fats account for about 33% of dietary lipid intake (14% of total energy intake). Although dairy products have a relatively low content of unsaturated fatty acids, such as linoleic and linolenic acids (because of hydrogenation of unsaturated fats in the rumen), the supply in milk of these so-called *essential fatty acids* may not be insignificant. The average daily intake of linoleic acid in Britain is about 10 g, although some people consume far less, and some far more, than this, while the average intake from dairy products is about 0.5 g per day. Milk fat is also an important source of the fat soluble vitamins, A, D, E and K, and it is particularly rich in vitamin A.

Despite such nutritional properties, there has been a growing concern in recent years that consumption of fats of animal origin may be a causal factor in the high incidence of heart disease which is evident in developed countries. To appreciate the force of this claim we need to make a brief digression into the field of pathology.

The most common cause of mortality from cardiovascular disease is *ischaemic heart disease* (IHD), in which the blood supply to heart muscle is reduced by occlusion of a coronary artery by a *thrombus* (blood clot). This occurs more readily when the artery has become narrowed by progressive *atherosclerosis*, an irregular thickening of the luminal wall caused by accumulation of *plaque*. The latter consists of smooth muscle cells, connective tissue and deposits of lipid, largely composed of cholesterol esters. The incidence of atherosclerosis appears to be influenced by the relative blood concentrations of *low density lipoproteins* (LDL) and *high density lipoproteins* (HDL) (see Figure 5.6), which are inversely related. Population surveys show that, in general, a high LDL:HDL ratio is associated with high IHD incidence. The concern over animal fat consumption stems from the observation that saturated fat (the predominant form of animal fat) tends to elevate blood concentrations of LDL. Polyunsaturated fats, on the other hand, tend to depress LDL concentrations.

Nevertheless, a sense of proportion is necessary when assessing the potential hazards of consuming milk fat. Not only do we need to weigh the disadvantages against the benefits, but proposed alternatives need to be investigated with the same rigour. For example, there is some evidence that consumption of large amounts of polyunsaturated fatty acids may adversely affect the clotting properties of blood.

The wisest counsel would seem to be one of moderation. For those with sedentary and/or stressful occupations, a modest milk intake, or consumption of low-fat milk, is advisable: and the same applies to those with a recognized genetic predisposition to cardiovascular disease. But for most people in developing countries the risks would appear to be insignificant.

3.5. Milk carbohydrates

Lactose is the principal and most distinctive carbohydrate of most milks (see category (b) in section 3.3); exceptions are those of some sea lions, in which it is absent or, as with monotremes, present at low concentration only. Lactose is a disaccharide composed of a *galactose* moiety linked to glucose (Figure 3.4), the full designation being O-β-D-galactopyranosyl-($1 \rightarrow 4$)-α-D-glucopyranose. Human milk contains the highest recorded concentration—about 70 $\mathrm{g\,l^{-1}}$, which is much higher than that in cow's milk.

Lactose is hydrolysed by the intestinal enzyme *lactase* to its constitutent monosaccharides, but it appears to have wider nutritional significance than simply providing a source of metabolic fuel. Because it is not readily absorbed across

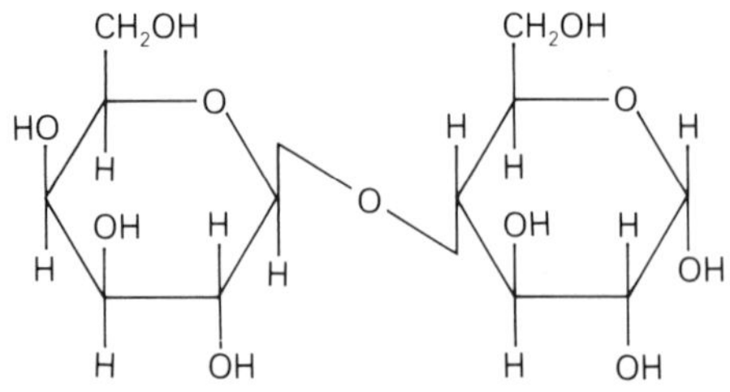

Figure 3.4. The lactose molecule.

the gut wall it is subject to significant microbial fermentation in the intestine, yielding lactic acid. The resulting fall in pH, by increasing the stability of calcium salts, aids their absorption into blood, an effect of considerable importance in view of the high calcium content of milk. The low pH also has the beneficial effect of inhibiting growth of basophilic (that is, chiefly proteolytic and putrifying) bacteria, while sustaining an acidophilic flora in the gut.

Milk contains a number of other carbohydrates, generally at much lower concentrations than lactose. These include the monosaccharides, glucose and galactose, and oligosaccharides containing, in addition to the latter, *fucose*, *N-acetylglucosamine* and *N-acetylneuraminic acid*. Milks of marsupials and monotremes contain a variety of such oligosaccharides, at higher concentration than that of lactose, in addition to monosaccharides.

3.6. Milk salts

These exist in milk in two forms: (a) as low molecular weight ions and complexes, that is *diffusible salts*; and (b) bound to protein, that is, *non-diffusible salts*. The most common cations are K, Na, Ca and Mg, and the most common anions, Cl, inorganic phosphate and proteins. Of these, K, Na and Cl are almost entirely diffusible. The major ions of the non-diffusible (colloidal) phase are Ca, Mg, inorganic phosphate and citrate, which are almost exclusively associated with the casein micelles. There are marked species differences in both the total concentrations of salts and in their partitioning between the colloidal and diffusible phases of milk.

3.7. Other milk constituents

Apart from the major classes of nutrients considered, milk contains a wide range of other substances present at low concentration, such as: cellular metabolites (for example, amino acids, urea, UDP glucose, vitamins), trace elements (for example, Zn, Cu, Mn), hormones (for example, prolactin, growth hormone) and 'growth factors' (which stimulate DNA synthesis and cell division). Some of these may have no particular significance for the growth and development of the young but others, for instance the growth factors, seem to have important nutritional consequences by stimulating intestinal growth. The fact that concentrations of such substances are low does not, of course, reflect on their importance, and in some cases (for example, trace elements), higher concentrations might well be detrimental. In terms of human nutrition, cow's milk is relatively deficient in iron and vitamins C and D.

Milk also contains cells, including polymorphonuclear leucocytes, macrophages, lymphocytes and degenerating mammary epithelial cells. Their concentration is highest in colostrum and just prior to involution. There is recent evidence that some of the lymphoid cells are immunologically reactive, and can become incorporated into the tissues of the neonate. Thus, immunological protection is provided by colostral secretion not only of Igs (see section 3.3.3), but also of immunocompetent cells.

3.8. General principles of milk composition and yield

Scientists faced with the bewildering array of milk compositions evident in different species have attempted to discover general principles which might explain the diversity. It seems hardly surprising that marked variations should exist when one considers the equally diverse range of conditions encountered by newborn mammals (see section 1.6.3).

One obvious factor influencing milk yield and composition is that relating to energy value. This has two aspects: (a) the relationship between the mother's metabolic rate and her resulting capacity for lactation; and (b) the energy requirements of the newborn for maintenance and growth.

Logarithmic plots of total energy in milk secreted per day at peak lactation *versus* maternal body weight (Figure 3.5) show that it is possible to distinguish two groups of species; group I corresponding to species with altricial young and group II to those with precocious young (see section 1.6.3).
For group I:

$$\mathrm{E_o} = 732 W_{\mathrm{M}}^{0.81}$$

where E_o is energy output in milk (kJ day^{-1}), and W_{M} is maternal mass (kg).

For group II

$$E_o = 347 W_{\mathrm{M}}^{0.73}$$

In neither case is the exponent significantly different from that in the, so called, Kleiber's Law:

$$M = kW_{\mathrm{M}}^{0.75}$$

which defines the relationship between basal metabolic rate (M) and body mass (W_{M}) for many species. Maternal body mass is an important determinant of milk energy output, but the maturity of the young at birth is also important, milk energy secreted in species with several young (altricial) being more than twice that in ungulates (hoofed animals) producing a single (precocious) neonate.

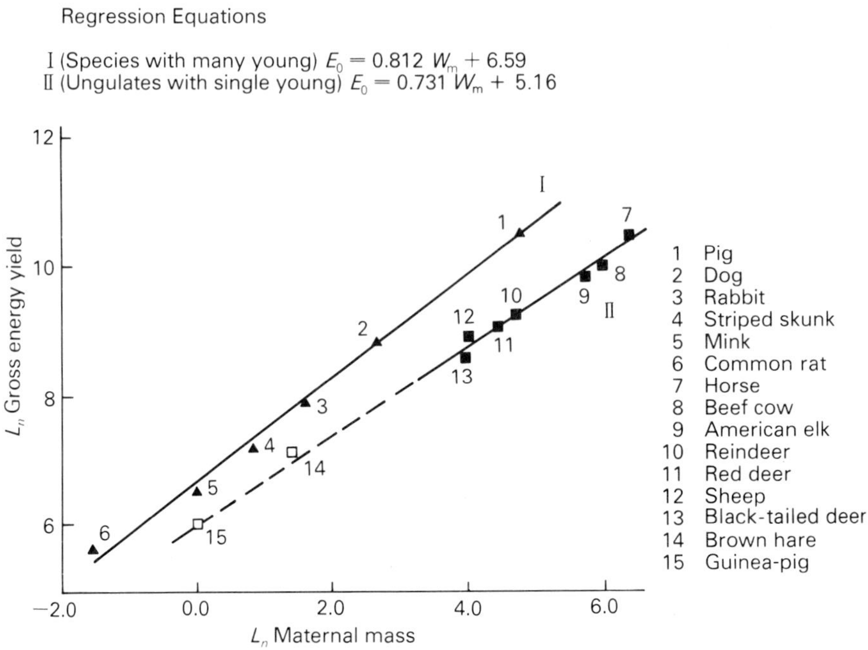

Figure 3.5. Relationship between energy output in milk and maternal mass after logarithmic transformation. The two regression equations are based on different data subsets. Species 14 and 15 were not included in the regression calculation, but fall close to the extrapolation of line II. (Adapted from Oftedal, 1984.)

When the energy requirements of litters are considered, an even closer relationship between neonate mass and E_o is obtained, that is, the standard error of the interspecies correlation is lower. It has been discovered that in suckling young metabolic rate is proportional to $W_N^{0.83}$ rather than $W_N^{0.75}$, where W_N is the average weight of an individual neonate. For several species, the following relationship applies:

$$E_o = 941 n W_N^{0.83}$$

where n is the number in the litter.

Some words of caution are, however, required. Calculations of total energy values of milk obviously involve determination of milk yield and composition, but for very few species can these be obtained with accuracy. Only for dairy ruminants, which are of docile temperament, possess capacious udder duct systems and show milk ejection in response to the abnormal stimuli of hand-milking or machine-milking, is determination of milk yield straightforward, and data for such species were deliberately omitted from the analyses described above. In fact, through genetic, nutritional and other improvements dairy cows have been, as it were, reduced to the metabolic size of dogs, and the world-record holder (yielding over 25 000 l in one year) to the metabolic size of a rat! Milk yields in species included in Figure 3.5 have thus been determined by techniques, such as estimation of litter weight gain due to suckling, which are sometimes of questionable accuracy. Moreover, analyses of milk composition rely on rather few data, some of which are also suspect. Thus, milk fat, a constituent with high energy value, shows considerable variation *during* the act of milking so that reliable energy values of milk depend on representative sampling of milk contained in the glands. Oxytocin, which is often administered to elicit milk ejection (see section 1.5), usually increases milk fat concentration. Finally, it must be remembered that the data discussed refer to peak lactation and do not necessarily apply to early or late lactation (see section 3.10).

Despite the inherent difficulties of such analyses, and perhaps because random errors tend to cancel out, the general trends relating to body size and energy value do appear to provide valuable insights. One practical application is the ability to predict nutrient requirements of rare species in captivity for which no data on milk yield or composition are yet available.

3.9. Principles determining milk composition

We have seen that the energy value of milk is related to body size, and hence metabolic rates, of both the mother and litter. Within that constraint, are there general principles governing the prevalence of different nutrients in the milks of different species and the total volumes of milk secreted? It appears that there are a number of such principles, the effects of which are, in some cases, in conflict. Milk composition for any particular species may thus be regarded as a compromise between competing pressures.

3.9.1. The isosmolarity principle

It appears to be a universal rule that milk is isosmotic with blood plasma, a feature of some importance because the maintenance of an osmotic gradient would entail energy expenditure. Since protein and fat are present in milk as large particles, osmotic pressure is chiefly determined by the carbohydrates and diffusible ions, and both within, and between, species there is generally an inverse relationship between lactose concentration and the sum of K and Na concentrations.

3.9.2. Ecological factors

It has been claimed that stable environmental conditions favour adoption of a reproductive strategy in which few young are born, but at an advanced stage of development: there is an intense competition between individuals so that the population is kept close to the carrying-capacity of the environment. This so-called *K-selection* contrasts with the *r-selection* exhibited by species occupying unstable environments, where multiple births of immature young, which show rapid *postnatal* development, is the rule. The two reproductive strategies, corresponding to the precocial and altricial categories already discussed (see section 1.6.3), are reflected in milks of distinctly different protein content. Species with precocious young (for example, primates and ungulates) secrete milk of lower protein concentration than those producing altricial young (such as rodents).

3.9.3. Relationship between protein and fat concentrations

With few exceptions, there is a positive correlation between protein and fat concentration of milk of different species (Figure 3.6a).

3.9.4. Relationship between fat, lactose and protein concentrations

There is a strong negative correlation between milk fat and lactose concentrations for most milks (Figure 3.6b) and, in view of section 3.9.3, also between milk protein and lactose concentrations.

3.9.5. Relationship between casein, calcium and phosphorus concentrations

Calcium and phosphorus concentrations are positively correlated, a fact presumably consequent on the prevalence of colloidal calcium phosphate in many milks. Moreover, the sum of calcium and phosphorus concentrations is positively correlated with casein concentration, suggesting that the 'casein–calcium phosphate' complex has a relatively constant stoichiometry.

3.9.6. Aquatic, arctic and desert species

Milks of these species show a common feature, very high fat concentration, despite the marked differences in their habitats. In the case of aquatic (for example, whales and seals) and arctic (for example, arctic fox) species, it is likely that the high fat content is important in facilitating subcutaneous fat deposition in the suckling young, thus reducing heat loss; moreover, the high energy value of fat (see section 3.4.1) serves to maintain the high metabolic rates demanded by low environmental temperatures. An alternative explanation, which also applies to desert species like the camel and ibex, is that the high fat content represents a means of conserving maternal water. The low water intake of desert mammals may be surpassed by that of phocid seals, which abstain from feeding and drinking during the several weeks of lactation, so that all water for milk secretion (and other purposes) must come from metabolism of body stores, chiefly in the form of fats.

3.9.7. Frequency of suckling

There are wide variations in suckling frequency. For example, mice suckle at approximately 20 min intervals (although the young are attached continuously to the teats), pigs hourly, rabbits once daily, tree shrews once every 2 days and northern fur seals about once a week. Despite the influence of other factors, there is a clear inverse relationship between the concentration of nutrients in milk and suckling frequency.

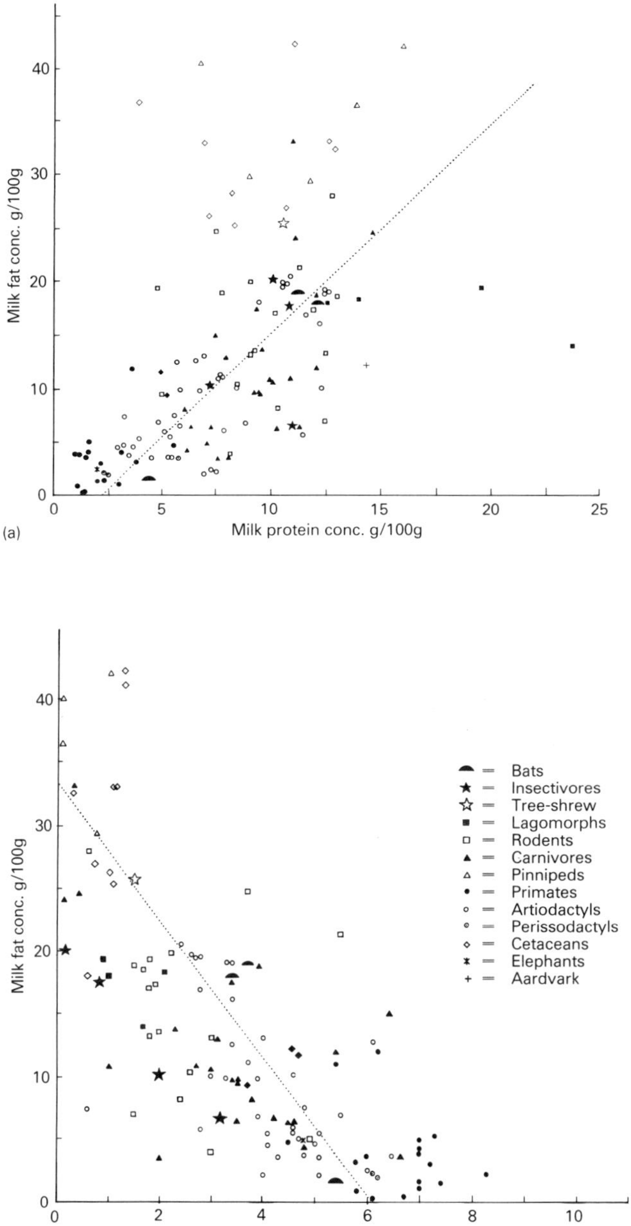

Figure 3.6. (*a*) Relationship between fat and protein concentrations in milks of several species. (For key to symbols see (*b*). (*b*) Relationship between fat and sugar (usually, lactose) concentrations in milks of different species.
(Source: Martin, 1984.)

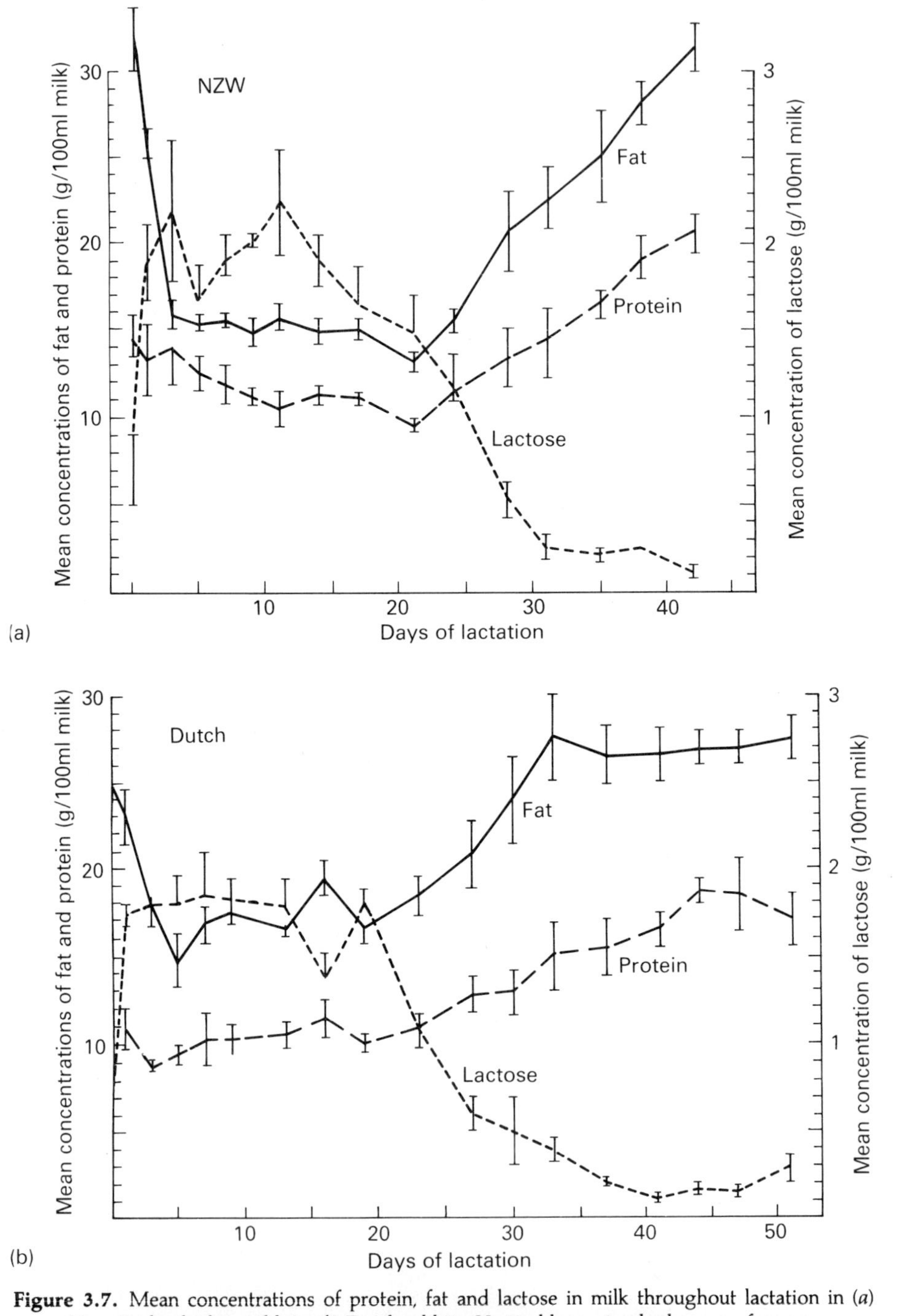

Figure 3.7. Mean concentrations of protein, fat and lactose in milk throughout lactation in (*a*) New Zealand white rabbits, (*b*) Dutch rabbits. Vertical bars, standard errors of means.
(Source: Cowie, 1969.)

3.10. Changes in milk composition throughout lactation

The data discussed above have been derived largely from measurements on milk collected at peak lactation, but it is important to realize that milk composition changes appreciably during the course of lactation (Figure 3.7). Very few longitudinal studies have been performed, but the wide divergence of patterns of change evident, even in two breeds of the same species, provide a salutary warning of the dangers of attempting to generalize from few data.

In cows, the high protein content which occurs immediately postpartum is largely accounted for by raised Ig concentrations, while the rising fat and protein concentrations concomitant with the falling lactose concentration indicate that the negative correlations noted for interspecies comparisons (see sections 3.9.3 and 3.9.4) also apply to individual animals.

Further reading

Butler, J. E. (1974). Immunoglobulins of the mammary secretions, in *Lactation* Vol. 3, pp. 217–55. eds Larson, B. L. and Smith, V. R. (New York and London: Academic Press.)

Casey, C. E. and Hambidge, K. M. (1983) The nutritional and immunologic significance of mammary secretions, in *Lactation: Physiology, Nutrition and Breast-feeding*, pp. 199–248, eds Neville, M. C. and Neifert, M. R. (New York and London: Plenum.)

Christie, W. W. (1983). The composition and structure of milk lipids, in *Developments in Dairy Chemistry* Vol. 2. pp. 1–35, ed. Fox, P. F. (London and New York: Applied Science Publishers.)

Davies, D. T., Holt, C. and Christie, W. W. (1983). The composition of milk, in *Biochemistry of Lactation*, pp. 1–11, ed. Mepham, T. B. (Amsterdam and New York: Elsevier.)

Gurr, M. I. (1983). The nutritional significance of lipids, in *Developments in Dairy Chemistry* Vol. 2, pp. 365–417, ed. Fox, P. F. (London and New York: Applied Science Publishers.)

Hambraeus, L. (1982). Nutritional aspects of milk proteins, in *Developments in Dairy Chemistry*. Vol. 1, pp. 289–313, ed. Fox, P. F. (London and New York: Applied Science Publishers.)

Hambraeus, L. (1985) Importance of milk proteins in human nutrition: physiological aspects, in *Milk Proteins '84*, pp. 63–79, eds Galesloot, T. E. and Tinbergen, B. J. Wageningen: Pudoc.

Hansen, R. G. (1974) Milk in human nutrition, in *Lactation*, Vol. 3, pp. 281–308, eds Larson, B. L. and Smith, V. R. (New York and London: Academic Press.)

Hayward, A. R. (1983). The immunology of breast milk, in *Lactation; Physiology, Nutrition and Breast-feeding*, pp. 249–70, eds Neville, M. C. and Neifert, M. R. (New York and London: Plenum).

Head, J. R. and Beer, A. E. (1978). The immunologic role of viable leukocytic cells in mammary exosecretions, in *Lactation*, Vol. 4, pp. 337–64. ed. Larson, B. L. (New York and London: Academic Press.).

Jelliffe, D. B. and Jelliffe, E. F. P. (1978). *Human Milk in the Modern World*. (Oxford and New York: Oxford University Press.)

Jenness R. (1974). The composition of milk, in *Lactation* Vol. 3, pp. 3–107, eds Larson, B. L. and Smith, V. R. (New York and London: Academic Press.)

Kon, S. K. (1972). *Milk and Milk Products in Human Nutrition*, 2nd edition (revised) (Rome Food and Agriculture Organization of the United Nations.)

Linzell, J. L. (1972). Milk yield, energy loss in milk, and mammary gland weight in different species. *Dairy Sci. Abstr.* **34,** 351–60.

Martin, R. D. (1984). Scaling effects and adaptive strategies in mammalian lactation. *Symp. Zool. Soc. Lond.* **51,** 87–117.

Oftedal, O. T. (1984). Milk composition, milk yield and energy output at peak lactation: a comparative review. *Symp. Zool. Soc. Lond.* **51,** 33–85.

Paige, D.M. (1982). Lactose digestion and milk intolerance in human populations, in *Animal Products in Human Nutrition*, pp. 373–83, eds Bietz D.C. and Hansen, R.G. (New York and London: Academic Press.)

Renner, E. (1983) *Milk and Dairy Products in Human Nutrition* (Munich GmbH, Volkswirtschaftlicher Verlag).

Swaisgood, H.E. (1982). Chemistry of milk protein, in *Developments in Dairy Chemistry*, Vol. 1, pp. 1–59, ed. Fox, P.F. (London and New York: Applied Science Publishers.)

Thompson, M.P. and Farrell, H.M. Jr. (1974). Genetic variants of the milk proteins, in *Lactation*, Vol. 3, pp. 109–34, eds Larson, B.L. and Smith, V.R. (New York and London: Academic Press.)

4 An Analysis of Lactation as a Productive System

4.1. Introduction

At this point in the book it will be helpful to pause in our account to consider some general principles.

We have seen that, in a metabolic sense, the mammary gland is a highly active organ, synthesizing and secreting large quantities of specific chemicals. We have noted that milk is stored in the gland following its secretion and is removed at intervals; and that the volume and persistence of milk production is crucially dependent on the intensity and frequency of suckling or milking. Before studying the finer details it is useful to survey the general features of the whole process.

Investigation of any system requires that it be isolated from factors (some of which might be unknown) which influence it. 'Isolation' does not necessarily imply physical separation, although this is clearly the case in biological studies carried out on organs or parts of organs *in vitro*. But at a conceptual level it is inevitable that a barrier is placed around the object of our investigations to more readily relate the occurrences within it. In erecting these conceptual barriers we distort the true situation because, in the words of the late Jacob Bronowski, 'the world is totally connected'. Scientific knowledge is thus, at best, an approximation to truth and its advance is at least partly attributable to the widening of the conceptual framework in which scientists operate.

4.2. Levels of study

There are several levels at which a phenomenon like lactation can be studied, ranging from the subcellular level, where observed events are expressed in biochemical terms, to the level of whole organisms in their physical and social environments, where behavioural activity is of major concern. Because of their training and particular skills most scientists concentrate their research on aspects of a single level. While this form of stratification has the indisputable benefits which derive from specialization, it runs counter to the current objective, which is to examine lactation as an integrated biological process. In pursuing this objective the attempt will be made to synthesize the principles which emerge from studies conducted at the different levels, minimizing the tendency to view them as closed systems. Nevertheless, since it is impossible to say everything at once, it is almost inevitable that, in discussing any particular topic, one or other level of study will predominate. In essence, the approach adopted corresponds to an elementary form of 'systems analysis', in that the complex of biological processes which constitute the phenomenon of lactation is resolved into a number of simpler elements. It is primarily an exercise in drawing logical deductions from the application to a particular system of generally accepted physiological and biochemical premisses.

4.3. Sites of metabolic conversion

It was noted in Chapter 3 that in all species studied the major milk components, fat, protein and carbohydrate, are organ-specific. They must therefore be produced within the mammary glands from materials delivered in the blood stream. Since these materials were originally ingested, we can construct a simple scheme outlining the sites of metabolic conversion (Figure 4.1). There are other possible routes by which material might, in theory, enter the mammary gland, for example, via the teat canal (a consideration discussed in section 3.3.3 in relation to the gland's immunological function) or from nerve endings; such routes are not, however, significant in the present context.

From this simple scheme (see Figure 4.1) we can deduce three fairly obvious, but important, statements:

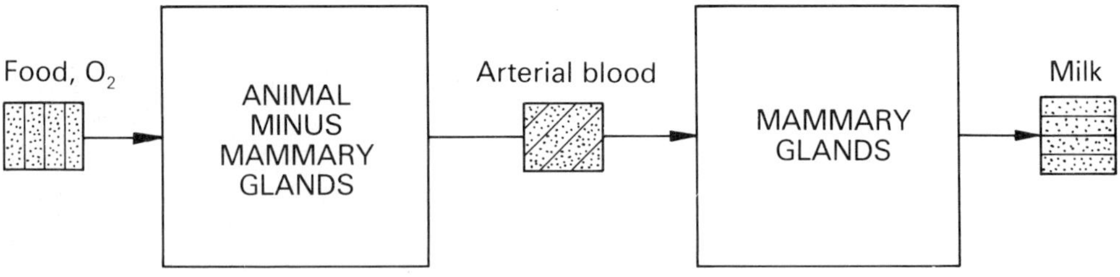

Figure 4.1. Sites of conversion of food into milk: a highly simplified scheme.

Milk produced is related to the amount and composition of food consumed.
Milk produced is related to the amount and composition of the blood supply to the mammary gland.
Since the volume of the gland is limited, the amount of milk produced is related to its rate of removal.

4.3.1. Nutrient partitioning

Next, let us consider the nature of the metabolic changes occurring within the lactating animal (Figure 4.2). Energy in various forms (for example, the potential energy in chemical bonds of ingested food and thermal energy in the environment) is used by the animal to grow, move, maintain temperature and secrete milk, while some is lost by excretory and evaporative processes. Clearly, a change in the amount and quality of the food intake is only one of several possible ways

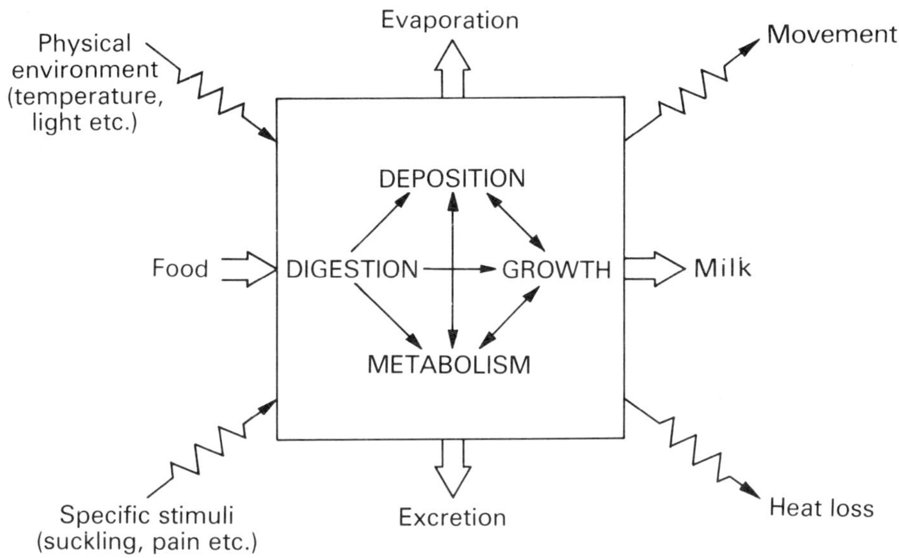

Figure 4.2. A summary of metabolic changes in a lactating animal, showing forms of energy input and output.

by which milk yield may be altered. The distribution of absorbed energy between various pathways (growth, deposition of body reserves, expenditure in muscular activity, etc.) may also play a significant role. Consider, for example, the body's utilization of glucose. Absorbed into the blood following the enzymatic hydrolysis of dietary polysaccharides, glucose serves as an important intermediate in many metabolic pathways. That produced in excess of current needs may be stored either as glycogen (for example, in the liver) or, following lipogenesis, in adipose tissue. With the onset of lactation a new channel of glucose utilization is created which inevitably affects its apportionment between the mammary gland and other tissues.

4.4. Mammary partitioning of absorbed substances

The same approach can be adopted in considering the ways in which the mammary glands utilize substances conveyed to them in arterial blood (Figure 4.3). As discussed (see section 2.2.4) there are several cell types in the gland. For present purposes we may group these into two classes: secretory (the mammary

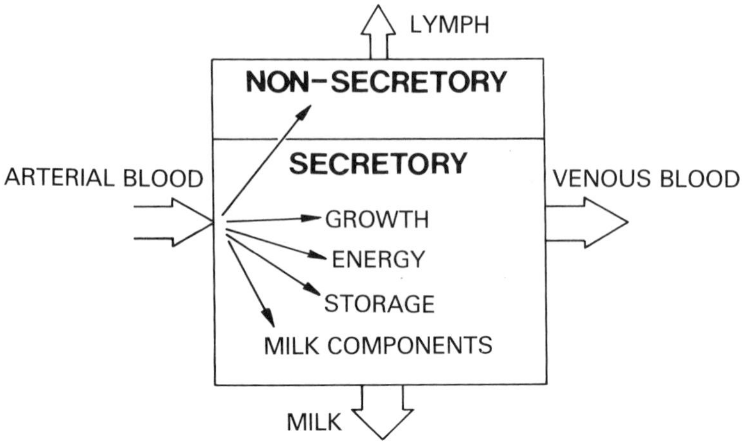

Figure 4.3. Substrate utilization by cells of a lactating mammary gland.

epithelium) and non-secretory (adipose, myoepithelial and connective tissue cells etc). Substances absorbed by the gland may thus be used either by secretory tissues, and as a result be incorporated into milk, or by non-secretory tissues, which have only an indirect bearing on milk yield. However, not all substances absorbed by the mammary epithelium necessarily appear in milk. Some, particularly during mammogenesis, are used in cell division and growth, some may be stored in the cell (for example, as glycogen), while others may be used in catabolic processes (for example, yielding energy in the form of ATP) of which the products leave the gland in venous blood. The way in which particular intracellular substances are distributed between the various pathways of utilization clearly has an important bearing on the amount and composition of the milk secreted.

There are three possible routes of exit for substances leaving the mammary gland (see Figure 4.3) (ignoring quantitatively insignificant routes like evaporation from the skin). Apart from secretion in milk, they may leave via the venous blood and in lymph. As far as is known, the latter is similar in composition to venous plasma, but of quantitatively negligible volume. While venous blood carries material not absorbed by the gland it also serves as an excretory route for substances produced in the gland (for example, CO_2).

From the foregoing we may make the following deductions. Milk produced by the mammary gland is related to:

the relative amounts of secretory and non-secretory tissues;
the manner in which absorbed substances are distributed between the alternative metabolic pathways in the secretory cells.

4.5. Mammary uptake of substrates

The mammary epithelium shows two types of activity. Firstly, it synthesizes those milk components defined as organ-specific (see section 3.3) from substrates absorbed from the blood; secondly, it acts as a selective barrier, in permitting the transfer of only certain substances from blood to milk. In each case the activity is determined by the rate of uptake.

Mammary uptake may be defined in a number of different ways, which are all, however, logically consistent. Thus, for any substance:

(a) Mammary uptake = quantity delivered in arterial blood − quantity leaving in venous blood.
(b) Mammary uptake = (concentration of substance in arterial blood- − concentration of substance in venous blood) × blood flow rate.

Also, for a substrate which is converted to a milk constituent the following equation applies in a stoichiometric sense (that is, where the *weight* only of material is considered and the milk constituent is a different substance from that absorbed):

(c) Mammary uptake = output of product in milk + lymph output + tissue storage

If the term 'output of product in milk' is altered to 'output in milk' the latter equation also applies to substances which pass unchanged from blood to milk (such as calcium).

The difference in concentration of a substance in arterial and mammary venous blood (*the arteriovenous difference,* more usually referred to as the *AV difference*) is relatively easy to determine for many substances in animals which have appropriate blood vessels close to the skin surface, thus facilitating blood sampling by hypodermic syringe (for example, lactating dairy ruminants; see Figure 2.6). Even in the absence of data on mammary blood flow rate (which it is also possible to obtain with relative ease in these ruminant animals; see section 4.5.1) the AV difference may be used to assess the efficiency of substrate uptake. This is usually defined as the *percentage extraction,* that is:

$$\text{percentage extraction} = \frac{\text{AV difference}}{\text{arterial concentration}} \times 100$$

4.5.1. *Quantitative considerations*

The above relationships may be clarified by reference to Figure 4.4. Consider the example of glucose, present in arterial blood at a concentration of x mg ml^{-1} and in the mammary venous blood at a concentration of y mg ml^{-1}. Its AV difference is therefore $(x-y)$mg ml^{-1}. If the rate of mammary blood flow (MBF) is R ml min^{-1}, then the uptake of glucose is given by:

$$\text{Uptake} = R\,(x-y)\ \text{mg min}^{-1} \text{ (from equation (b) section 4.5)}$$

The efficiency of uptake is represented by the percentage extraction:

$$\text{percentage extraction} = \frac{(x-y)}{x} \times 100\%$$

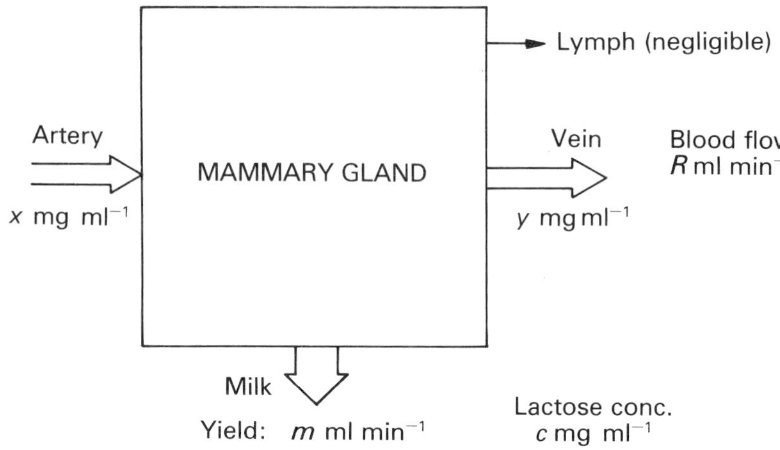

Figure 4.4. Quantification of substrate uptake by a lactating mammary gland using the AV difference method. (For explanation, see text.)

Several methods have been employed to determine MBF. For routine use in ruminants the most satisfactory is thermodilution, in which saline at ambient temperature is infused slowly into the milk vein (V in Figure 2.6, page 21) close to the udder and the drop in temperature of the blood a few centimeters downstream is measured with a needle-mounted thermistor. The blood flow is inversely proportional to the temperature drop.

In Chapter 5 we shall consider the role of glucose as a precursor of lactose in milk. But irrespective of specific biochemical considerations, the data already discussed permit us to assess the potential contribution of glucose uptake to lactose output. Thus, if the milk yield is m ml min^{-1} and its lactose concentration is c mg ml^{-1}, then the output of lactose $= cm$ mg min^{-1}. It follows that when:

$$R(x-y) \geqslant cm$$

enough glucose is absorbed by the gland to account for all the lactose secreted, assuming lymph flow and tissue storage to be negligible (compare equation (c) section 4.5).

Of course, this stoichiometric approach may prove to be too simplistic. Glucose may be required for purposes in the mammary cell which preclude its utilization for lactose synthesis in the way suggested: such questions are discussed in Chapter 5. Nevertheless, the combination of AV sampling, MBF measurement and chemical analysis provides a valuable insight into the quantitative relationship between precursors and products, which it has proved much easier to obtain for the mammary gland than for most other organs.

There is, however, one very important proviso in the use of such measurements to quantitate precursor–product relationships: that 'steady-state' conditions apply. Specifically, reliable estimates of the extent to which there is balance between substrates uptake and product output depend on constancy of (a) MBF, (b) arterial substrate concentration and (c) tissue uptake of the substrate. Representative sampling of the venous blood is also a prerequisite. Accuracy thus demands that adequate checks are made to assess variability, while animals used must be free from external sources of disturbance.

4.6 Mammary blood flow

The dependence of mammary substrate uptake on MBF rate was indicated by equation (b) (section 4.5). It is useful to redefine that relationship in different terms to highlight other factors involved. Thus:

$$\text{Mammary uptake} = \text{extraction } \% \times \text{arterial concentration} \times \text{MBF}$$
$$\text{Mammary uptake} = \text{extraction } \% \times \text{supply rate to gland}$$

The amount absorbed (that is, the uptake) is thus a measure of the efficiency with which the gland retains substances supplied to it in the blood. The latter amount can be increased by increased MBF, by increased arterial concentrations, or by both of these occurring simultaneously.

The interdependence of MBF and substrate uptake is reflected in the positive correlation observed between blood flow and milk yield in lactating cows and goats. In both species, in established lactation, approximately 500 volumes of blood perfuse the mammary glands for every one volume of milk secreted. In late lactation this ratio increases to over 1000:1 (and obviously to infinity when the animal is 'dry', or producing no milk), representing a decreased efficiency of milk production in these terms.

4.6.1. Determinants of mammary blood flow

Mammary blood flow (as that through any organ) may be considered dependent on two factors: the cardiac output (that is, the output of the heart per unit time) and the fraction of the latter which is distributed to the glands. Other things being equal, an increased cardiac output will increase flow through all organs of the body. This is one mechanism by which supply to the mammary glands is regulated, output increasing when animals begin to lactate.

The distribution of blood between different organs of the body is subject to haemodynamic factors. For any organ:

$$\text{blood flow} = \frac{\text{arterial pressure} - \text{organ venous pressure}}{\text{resistance of organ}}$$

It follows that for a given arterial pressure (itself dependent on cardiac output and the total peripheral resistance) higher flows through any particular organ will be achieved by a decreased resistance in the organ, a decreased venous pressure, or both processes together. Haemodynamic resistance to flow through the organ's vasculature is thus a major flow determinant. This resistance depends on three factors:

viscosity of the blood,
length of blood vessels in the organ, and
the radius of the blood vessels.

Since the first two factors may be regarded as essentially constant for the mammary gland, vascular radius is the major determinant of flow. Arterioles are the principal site of vascular resistance, capillary resistance being considerably less. Moreover, only a fraction of the capillary bed might be perfused at any one time, so that flow may be increased by the opening up of previously collapsed vessels. Since resistance is inversely proportional to the radius[4] of any vessel (*Poiseuille's Law*), modest changes in vessel diameter have profound effects on resistance (for example, halving the diameter increases resistance 16 times).

Factors leading to an increased MBF are summarized in Figure 4.5. Changes

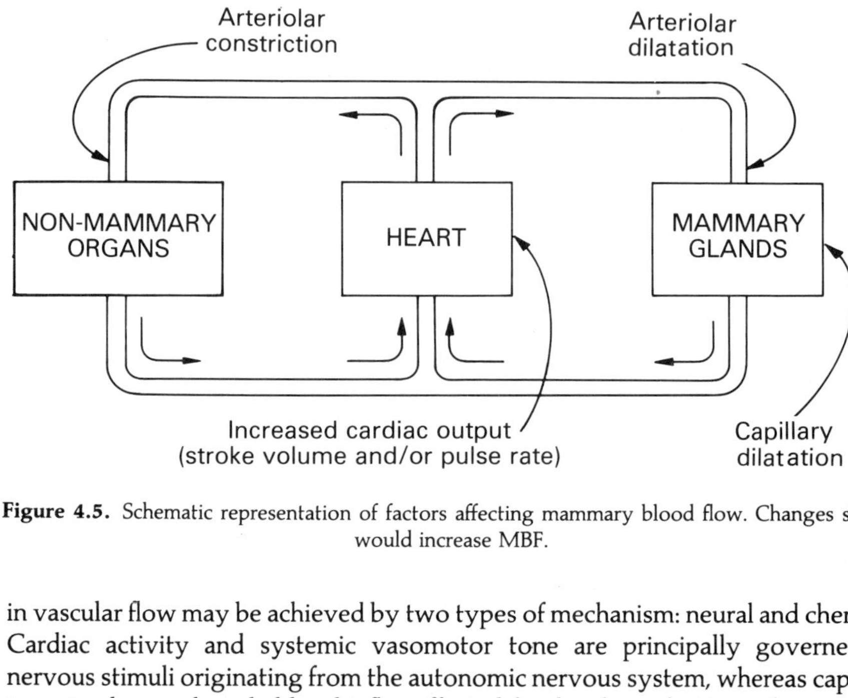

Figure 4.5. Schematic representation of factors affecting mammary blood flow. Changes shown would increase MBF.

in vascular flow may be achieved by two types of mechanism: neural and chemical. Cardiac activity and systemic vasomotor tone are principally governed by nervous stimuli originating from the autonomic nervous system, whereas capillary tone is also, and probably chiefly, affected by local production of vasoactive chemicals in the mammary gland.

In terms of substrate supply for milk synthesis, only that fraction of the total MBF perfusing secretory tissue is relevant. This fraction might vary, not only with stage of lactation, but also for example, with changing environmental conditions. Thus, a high temperature will result in an increased flow through the skin. Even blood apparently perfusing secretory tissue may, in fact, be of little value in supplying substrates because mammary tissue contains arteriovenous shunts (see section 2.2.5) by which the capillary beds may be bypassed.

Since blood is the sole route by which material appearing in milk is supplied to the gland, as well as an essential means by which its activity is controlled (see section 4.13), it is clear that cardiovascular mechanisms, at both systemic and local levels, are crucial in regulating milk yield and composition.

4.7. The interstitial fluid compartment

Although blood is the source of substances absorbed by the secretory cells, an extra compartment is interposed between these two: the interstitial fluid (ISF), which bathes the cells (see Figure 1.1). (The 'ECF' comprises both ISF and blood plasma, in a volume ratio of about 4:1, the chief compositional difference being the low concentration of plasma proteins in ISF.) The composition of ISF is dependent on ultrafiltration and diffusion of substances from blood plasma (and hence on the permeability of the capillary wall) coupled with the effects of substrate uptake by the secretory epithelium.

Substrate uptake by secretory cells thus occurs at the ISF/basal membrane interface. Where simple diffusion is involved only the concentration gradient between ISF and intracellular fluid (ICF) is important. But for many substrates, uptake is achieved by binding to specific *carriers* (see section 5.3.2) prior to transport into the ICF. In such cases, the affinity of the substrate for the carrier, and the concentration of carrier molecules in the membrane, are also important determinants of uptake.

4.8. Arterial substrate concentrations

We have seen that the supply of substrates to the mammary gland is the product of MBF and arterial concentrations (see section 4.6). These concentrations are the result of the relative rates of inflow of the substances into the blood from all sources (intestinal tract, liver, muscle, etc.) and their removal from it by the body's organs or by excretion. Hormones which have major effects on metabolic processes, such as insulin and growth hormone, are certain to be important in regulating blood concentrations of substrates over shorter periods of time, as is food intake over longer periods. The adjustment of body metabolism to meet the changed priorities of lactating animals is an example of a phenomenon which has been termed *homeorhesis* (by analogy with homeostasis); it is discussed in Chapter 10.

4.9. Roles of the duct system

In the majority of mammalian species (marsupials are a possible exception) milk is stored in the duct system of the mammary gland because suckling is an intermittent activity. A storage role is an unusual feature of exocrine glands and it raises the question of how the rate of milk secretion is affected by its accumulation in a closed cavity. For current purposes the gland may be represented as a 'two-compartment' system (Figure 4.6). Secretion of milk into the duct system will raise the intraductal pressure, but there is not a linear relationship between volume and pressure because relaxation of smooth muscle in the ductal walls allows the accommodation of milk at low pressure until the elastic limit of the tissues is approached, when pressure rises more steeply. The situation has been

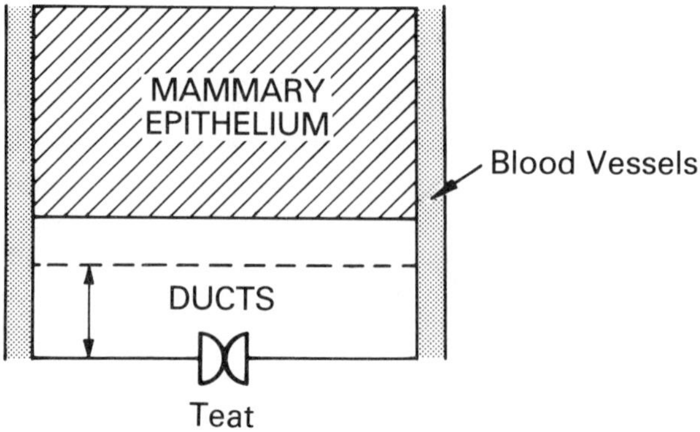

Figure 4.6. A simple, 'two-compartment', model of milk storage space in mammary tissue.

compared to the inflation of a balloon inside a slightly elastic string bag. This 'two-compartment' model (see Figure 4.6) is, however, certain to be a gross simplification. Milk is secreted into the hollow lumina of the alveoli (see section 2.2.2) and only gains access to the large ducts and cisterns in the gland via fine-bore ducts draining the alveoli. For flow to occur the pressure in the alveoli will thus need to be greater than that in the cisterns in order to overcome the resistance to passage through the small ducts. Moreover, in animals with pendulous glands, such as cows, the pressure in the large ducts will depend on the position at which it is measured, since it will partly depend on the mass of milk vertically above it. Thus, for a number of reasons, intraductal pressure is most unlikely ever to be uniform throughout the system.

4.9.1. Effect of milk accumulation

It is clear that, in the absence of milk removal, its accumulation in the duct system will eventually inhibit secretion because, if for no other reason, the rising intraductal pressure will require increasingly greater energy expenditure in transferring milk components across the apical membranes of the secretory cells. The raised pressure may also, by transmission through the glandular tissue, affect MBF (see Figure 4.6). Thus a high pressure may cause closure of blood capillaries and inhibit milk secretion by reducing substrate supply. But secretion rate might also be inhibited by more subtle, chemical means. Although the composition of milk removed from glands is essentially constant, the possibility exists that there are local variations in the concentrations of specific components in different regions of the duct system. This is known to be the case for milk fat, which, probably because the large fat globules (see section 3.4) are impeded in their passage through the fine-bore ducts, is present at higher concentrations in milk removed at the end of a milking than in that removed at the beginning. Elevated concentrations of fat (and of fat soluble compounds), and possibly of other milk components, might thus reduce milk synthesis by the well-known phenomenon of *end-product inhibition* (see section 4.11). An alternative mechanism, operating on the same principle, would involve secretion of a specific chemical inhibitor (not normally recognized as a milk constituent) in small but effective amounts. A third possibility is that substances with an inhibitory action might be formed in the duct system during the period of storage, by modification of normal milk constituents (for example, by enzymatic proteolysis or lipolysis). Enzymes capable, in theory, of effecting such changes are known to be present in milk.

4.9.2. *Permeability of the duct epithelium*

In certain organs which store their secretory product (for example, salivary glands), extensive reabsorption into blood occurs. Does this happen in the mammary gland? A neat and effective way of answering this question has been employed in the goat, in which by inflating a pneumatic cuff around the base of the teat an isolated 'teat pouch', lined with duct epithelium, may be formed

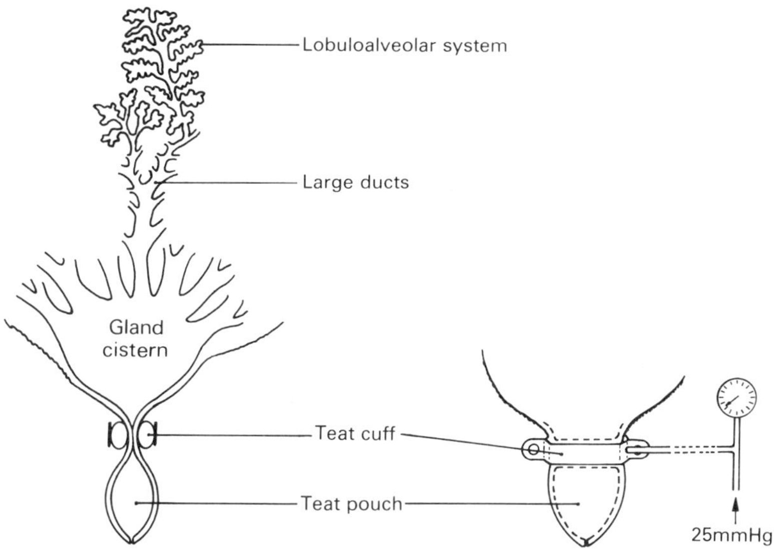

Figure 4.7. Formation of 'teat pouch' by use of a pneumatic cuff positioned at the base of the teat of a goat mammary gland. (Source: Linzell and Peaker, 1971.)

(Figure 4.7). The pressure in the cuff is sufficiently high to prevent passage of fluid into or out of the teat cavity, but not high enough to impede blood flow. By introducing radioactively labelled substances into the pouch through the teat canal and removing them some time later it has been shown that, while water equilibrates readily with blood, other milk components (such as Na, K, Cl, lactose) do not. Other evidence, including the morphological uniformity of the epithelial cells lining the duct system, suggests that the ducts as a whole are impermeable to the major milk components so that milk composition is essentially determined by the activity of the secretory epithelium. In addition to water, certain other substances which are lipid soluble or un-ionized (such as urea and carbonic acid) can, however, cross the ductal epithelium, suggesting that these cells constitute a *blood–milk* barrier of similar nature to the *blood–brain* barrier.

4.9.3. *Integrity of the mammary epithelium*

In the model proposed above (see Figure 4.6) the mammary epithelium was described as a 'compartment', the contents of which are physically separated from ISF on the 'blood' side and milk on the 'ductal' side (Figure 4.8a). If however, the effectiveness of the mammary epithelium as a physical barrier is imparied by weakening of the adhesive forces between adjacent cells (Figure 4.8b) equilibration

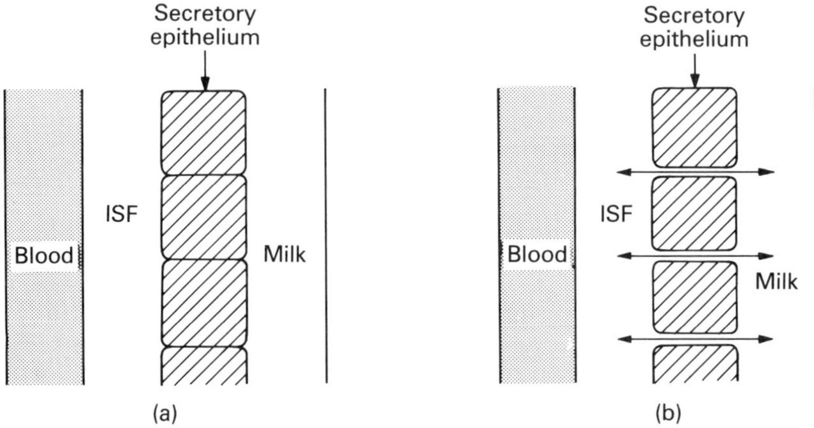

Figure 4.8. Effectiveness of the secretory epithelium as a physical barrier. (*a*) Tight junctions are competent and no fluid passes between cells. (*b*) Tight junctions are 'leaky', leading to equilibration of ISF and milk.

of milk and ISF will tend to occur (arrows). This condition may occur, to greater or lesser degree, in disease states like *mastitis* (inflammation of the gland) or in other conditions when the junctional complexes joining neighbouring cells (see section 2.3.2) are disrupted by stretching (for example, when the intraalveolar pressure is high due to excessive accumulation of milk). The route by which substances pass between cells is referred to as the *paracellular route* to distinguish it from the normal *transcellular route,*—through the cells.

4.10. Determinants of secretory potential

So far in this chapter we have been concerned largely with the metabolic routes by which ingested material is channelled to and through the mammary gland. In this and the following sections attention is turned to concepts of 'control'. As for any dynamic system, it is possible to identify several phases of productive activity in the lactation process. In essence, these may be listed as:

starting/increasing/decreasing/stopping

but such brevity obscures the true complexity of the process. For example, factors controlling secretion of the many different substances in milk are likely to be distinct but interactive, and they may not be the same at all stages of lactation or in all species. Again, factors operating to reduce secretion in a disease state will differ from those obtaining at mammary involution subsequent to normal regression.

Before considering control mechanisms it is useful to establish two simple principles relating to milk yield:

(a) Milk yield = potential yield × fraction realized

This is, of course, a monumental truism, but stating it emphasizes that for any individual animal there is an upper limit to milk yield, which is determined by its genetic constitution. However, since it is almost impossibly difficult to determine genetic potential by independent means, attempts to increase yields by non-genetic means will doubtless continue to be made, especially in dairy animals. Genetic improvement through selective breeding has long been an effective strategy for raising milk yields, but it seems likely that the increasing use of techniques of 'genetic engineering' will soon provide an alternative means of exploiting this potential (see section 11.4).

Milk secreted may also be regarded as the product of two other factors, which may be called an *extensity* factor (that is, the number of mammary secretory cells) and an *intensity* factor (that is, the average amount of milk secreted per cell).

(b) Milk yield = cell number × mean yield per cell

The amount of DNA in a mammary gland has been taken as an index of the number of cells (on the assumption that the DNA per cell is constant). Moreover, the RNA:DNA ratio for mammary tissue has been regarded as an index of secretory activity per cell. Neither of these indices is strictly valid, for example, in the former case because it does not distinguish between secretory and non-secretory cells, but with suitable correction for obvious defects they provide a useful guide to the extensity and intensity factors, respectively.

Increases in yield may thus, in theory, be achieved by increased cell numbers (increased DNA) or by increased secretory activity per cell (increased RNA:DNA ratio) or by both processes occurring together. By the same token, the same yield may be produced by a large, low activity gland as by a small, high activity gland. The concept of intensity and extensity factors might usefully be applied in other contexts. For example, the rate of protein synthesis is the product of ribosome number and their average translational rate.

4.11. Control of synthetic rates

For any specific component of milk, the rate of synthesis (as in any reaction sequence) is determined by the slowest step in the synthetic pathway. For example, in the formation of D from A (Figure 4.9a) via intermediates B and C,

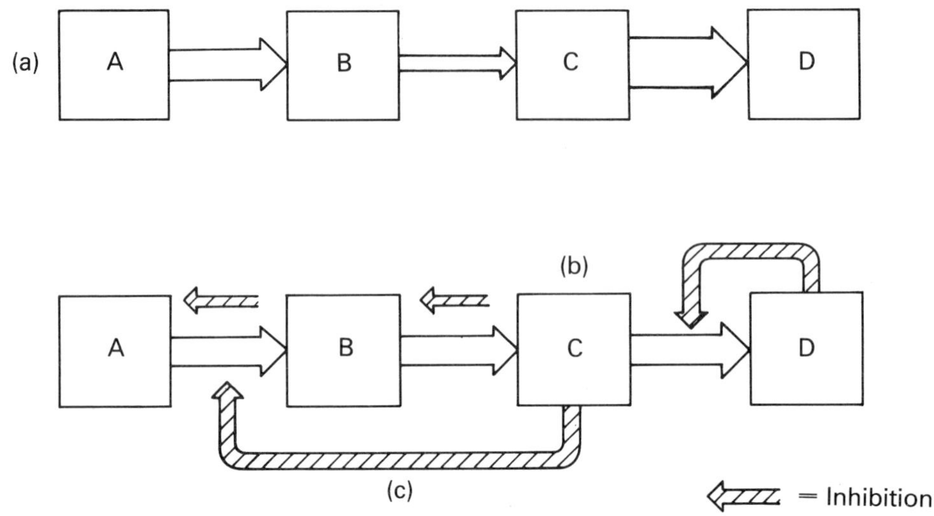

Figure 4.9. Rate-limitation of product synthesis. Width of arrows in (*a*) indicates potential rates of reaction. (For explanation, see text.)

formation of C is rate-limiting. In some cases accumulation of the product (b) may inhibit further synthesis (end-product inhibition) or one of the intermediates may specifically suppress one of the earlier steps in the sequence (c). Where the products of two pathways react to form a milk component the rate of formation of the latter will be limited by the slower pathway.

The *net* rate of synthesis or, what may amount to the same thing, the amount of a substance available for secretion from the cell, is dependent on the relative

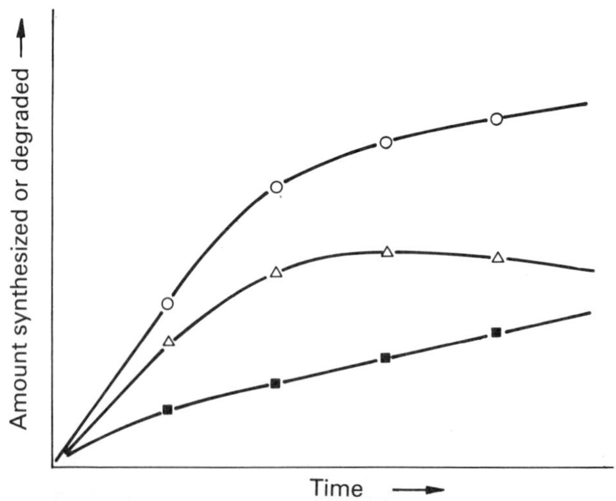

Figure 4.10. Rates of initial synthesis (o), degradation (●) and net synthesis (△). (For explanation, see text.)

rates of its synthesis and degradation (Figure 4.10). The possibility of degradation is underlined by the presence in secretory cells, even in early lactation, of lysosomes. The balance between synthesis and degradation is also an important factor in relation to the activity of the cellular 'machinery', for example, the enzymes and other proteins involved in transporting substances across membranes.

4.12. Sites of control in secretory cells

For any individual secretory cell, the amount of milk secreted may be considered to be subject to three factors:

- the concentration of metabolic precursors of milk components at the appropriate intracellular sites of synthesis;
- the activity of the relevant synthetic machinery (enzymes);
- the efficacy of intracellular and transmembrane translocation processes for the synthetic products.

In practice, all these factors are interrelated and adoption of this scheme should be viewed solely as a device to aid discussion.

Intracellular substrate concentrations are the result of two types of process: (a) translocation across various membrane barriers, and (b) intracellular synthesis. Thus, substrates present in the ISF (see section 4.7), derived from those delivered in the blood, must first penetrate the basal membrane of the secretory cell (and subsequently in some cases other, intracellular, membranes) before reaching their sites of metabolism. Other substrates are themselves synthesized within the cell from absorbed material before incorporation into milk components (for example, see section 5.3.1). Concentrations of substrates are likely to be quite different in different compartments of the cell so that the term 'intracellular concentration' is of doubtful biochemical value.

Once substrates have accumulated at the appropriate intracellular sites, subsequent rates of synthesis (for example, polymerization of amino acids to form proteins) depend on the activity of enzymes (subject to the presence of activating ions, co-factors etc.), a supply of energy (for example, in the form of ATP) and the intracellular translocation process which removes the reaction products from their sites of synthesis. The latter process may be presumed to be important not only because it is a prerequisite of secretion of the products across the apical membrane, but also because their accumulation at the sites of synthesis would tend to inhibit further synthesis.

4.13. Agents controlling secretory activity

Secretory activity of the mammary gland is a phasic process. It only occurs at specific times, such as following parturition or, more unusually, in response to hormonal changes consequent, for example, on a sterile mating or diseased state of the endocrine system. It follows that secretion is due to an interaction between structures in the secretory cells and stimuli which originate outside them. What, in theory, are the possible routes by which such stimuli could act on the cells?

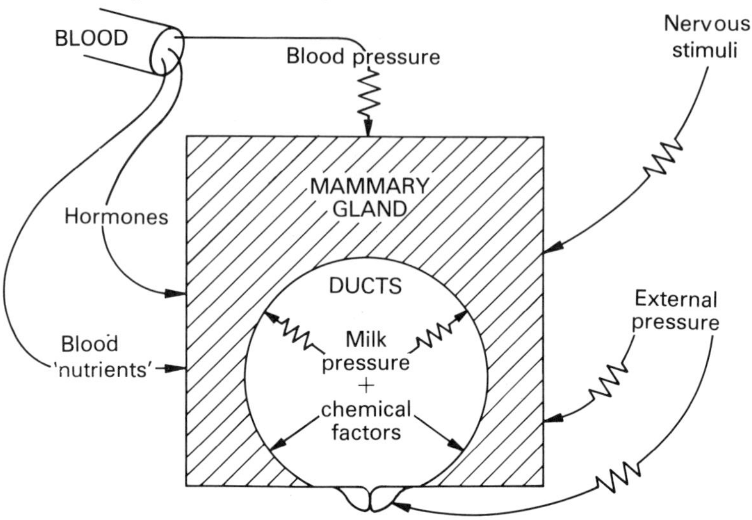

Figure 4.11. The forms of energy impinging on the mammary gland.

Figure 4.11 summarizes the different routes by which energy, in various forms, impinges on the mammary gland. Essentially, it is of three kinds:

chemical: hormones; blood 'nutrients', such as glucose; substances in milk within the gland;
physical pressure: blood pressure; that due to fluid in the duct system; that exerted on the outside of the gland;
electrical (nervous stimulation)

Let us consider how these agents might operate.

4.13.1. Chemical factors

Hormones exert a dominant and sensitive control over all aspects of mammary function, as will be discussed in Chapters 7, 8, 9 and 10. The principal systemic hormones and the glands from which they are secreted are listed in Table 4.1. In addition to these factors, which are transported in the blood, a number of other hormones may indirectly affect mammary function. These, so called *paracrine secretions* (Table 4.2), reach their target cells by *diffusion*, so that their direct effects are generally local: however, by modulating release of systemic hormones (for example, somatostatin inhibits GH secretion) they may exert a much wider influence. For present purposes we need only note that hormones act by modifying cell metabolism. This may be achieved by processes such as alteration of the permeability of cell membranes or stimulation of genes in the nucleus.

The concentration of non-hormone substances in blood (glucose, fatty acids, K^+ ions etc.) may alter metabolic rates simply through the 'law of mass action',

Table 4.1. Systemic hormones affecting the lactation process, their glands of origin and chemical nature.

Endocrine gland		*Hormone*	*Chemical structure*
Pituitary gland	Anterior lobe	Prolactin	Peptide
		Growth hormone (GH)	
		Adrenocorticotrophic hormone (ACTH)	
		Thyroid stimulating hormone (TSH)	Glycopeptide
		Luteinizing hormone (LH)	
		Follicle stimulating hormone (FSH)	
	Posterior lobe	Oxytocin	Peptide
		Vasopressin (ADH)	
Thyroid gland		Thyroxine (T_4)	Amino acid derivative
		Triiodothyronine (T_3)	
		Calcitonin	Peptide
Parathyroid gland		Parathormone (PTH)	Peptide
Pancreas		Insulin	Peptide
		Glucagon	
		Somatostatin	
Adrenal gland	Cortex	Corticosteroids	Steroid
	Medulla	Adrenaline	Catecholamine
Ovary		Oestrogen	Steroid
		Progesterone	
Fetoplacental unit		Placental lactogen	Peptide
		Sex steroids + other hormones	Discussed in Chapter 7
Mammary gland		Oestrogen, progesterone and other factors	

but they may also be important in other ways. Thus, they may promote uptake of other substrates by co-transport effects or regulate reaction rates by provision of co-factors or appropriate ions. Many of their effects may also be mediated by systemic hormones.

As noted (see section 4.9.1), normal constituents of milk or those which may be formed during storage in the duct system may also influence further secretion.

4.13.2. Effects of pressure

Blood pressure is an important factor determining MBF and hence the delivery rate of substrates and hormones to the mammary cells (see section 4.6).

External pressure *per se* would seem unlikely to have a specific effect on the

Table 4.2. Some peptide paracrine secretions which may affect the lactation process, and their distribution in tissues of the brain and gastrointestinal tract.

Peptide hormone	Hypothalamus	Pituitary Gland	Brain*	Gastro-intestinal tract
Substance P	+	+		+
Beta-Endorphin	+	+	+	+
Enkephalin	+	+	+	+
Somatostatin	+		+	+
Cholecystokinin (CCK)	+		+	+
Insulin			+	+
Secretin			+	+
Gastrin	+		+	+
Vasoactive intestinal peptide (VIP)	+		+	+

*excluding hypothalamus

secretory cells under normal conditions, although when excessively high it might constitute a significant resistance to milk flow through the duct system and to MBF (see section 4.9.1). The more subtle pressure stimuli detected by receptors in the teat have their effects via neuroendocrine reflexes, though in some species myoepithelial cell contraction might also occur in response to abrupt physical pressure exerted on the gland (see section 8.2.8).

4.13.3. *Neural effects*

Efferent nerves (that is, those carrying impulses to the gland, in contrast to the afferent nerves which are involved in the milk ejection reflex; see section 8.2.4) are largely confined to innervation of blood vessels and the musculature of the teat sphincter. Changes in vascular tone elicited by the former would clearly produce changes in MBF (and hence substrate delivery) via resistance effects (see section 4.6.1).

4.13.4. *Control of incremental and decremental phases*

It was noted in section 4.10 that yield is the product of cell numbers and secretory activity per cell. We may thus deduce that one or more of the above agents (see section 4.13) is responsible for stimulating the incremental phases of milk secretion (starting and increasing yield), either by increasing cell numbers or by increasing activity per cell, or possibly by both processes. However, while these agents may also be important in controlling the decremental phases (decreasing and stopping secretion), in these cases purely intramammary factors, like cell death, might also play an important part.

4.14. The cell as a factory

An analogy has often been drawn between the activities of a cell and those, more easily examined, in a typical factory. Since the nature of 'explanation' may often correspond to a process of drawing analogies between unfamiliar and familiar, and thus apparently comprehensible, systems, it is a useful exercise to consider lactation in this context. Indeed, since the mammary gland is so obviously a productive system the comparison would seem particularly apposite.

Raw materials derived from the body's resources (liver glycogen might be equated with mineral reserves in the earth's crust) are conveyed by a transport system (blood = lorries) to the basal membranes of the secretory cells (= factory gates). Carrier systems (= forklift trucks) convey them into the cell (factory) where they are distributed to the appropriate assembly points (for example, ribosomes = work benches). Through the agency of enzymes (tools) and energy (manpower), and with appropriate co-factors (mugs of tea?), assembly is conducted according to a blueprint (mRNA molecules). The enterprise is subject to managerial directives (central nervous system), which are channelled through supervisors (hormones). The process is very sensitive to demand (the suckling stimulus) and if alternative products prove more attractive (as at weaning) production may come to a halt (involution).

Such analogies should obviously not be pushed too far. Their value lies in encouraging one to think about problems in ways which might suggest new approaches to their solutions.

4.15. Conclusion

This chapter has been concerned with principles and possibilities. Except in a few cases where the mammary gland shows unique features (such as the duct system) the discussion has been concerned with the application of general physiological and biochemical concepts to lactation as a productive system.

In the chapters which follow there is a greater emphasis on what *has* been discovered about how milk is secreted and it will become evident to the reader that not all of the possibilities discussed in this chapter have been investigated. Some may appear, in the light of research findings, to be much more important than others. And it would defy all precedents if future research did not reveal new principles and other 'possibilities', as yet unforeseen.

Further reading

Barry, J. M. (1964). A quantitative balance between substrates and metabolic products of the mammary gland, *Biol. Rev.* **39,** 194–213.

Fleet, I. R. and Mepham, T. B. (1983). Physiological methods used in the study of mammary substrate utilization in ruminants, in *Biochemistry of Lactation,* pp. 469–91, ed. Mepham, T. B. (Amsterdam and New York: Elsevier.)

Linzell, J. L. (1971). Techniques for measuring nutrient uptake by the mammary glands, in *Lactation,* pp. 261–79, ed. Falconer, I. R. (London: Butterworths.)
Linzell, J. L. (1974). Mammary blood flow and methods of identifying and measuring precursors of milk, in *Lactation,* Vol. 1, pp. 143–225, eds Larson, B. L. and Smith, V. R. (New York and London: Academic Press.)
Vernon, R. G. and Peaker, M. (1983). The regulation of nutrient utilization: basic principles and mechanisms, in *Nutritional Physiology of Farm Animals,* pp. 41–113, eds Rook, J. A. F. and Thomas, P. C. (London and New York: Longman.)
Waddington, C. H. (1977). *Tools for Thought.* (St Albans: Granada.)

5 Biosynthesis of Milk

5.1. Introduction

In his famous book *Animal Chemistry*, first published in 1842, the renowned German chemist, Justus von Liebig, declared '... milk contains only one nitrogenous substance, known under the name of *casein*'; ('casein' is derived from the Latin *caseus*, meaning cheese). Liebig believed that this protein was identical to the plant protein *legumin*, present at high concentrations in leguminous plants. Milk protein secretion thus consisted of a process of absorption of this 'vegetable casein' into blood, followed by its filtration into the ducts of the mammary gland. The alleged Chinese practice of making a 'real' cheese from peas was cited in support of the theory.

The theory became untenable when use of improved analytical procedures established that casein is only present in milk; it was soon replaced by other theories which sought to explain the organ-specifity of the milk components. For example, in the 1880s it was proposed that albumin from the blood combined with 'nuclein' from mammary cell nuclei to form the 'nucleoprotein', casein: we now know that the albumin molecule is much larger than the casein molecule. Another suggestion was that albumin from blood was secreted into the alveolar lumina, where it suffered enzymatic degradation to yield casein. It was only when the polypeptide structure of proteins was established at the turn of the century, largely through the efforts of Emil Fischer, that it became evident that protein synthesis entails assembly of molecules *de novo* from individual amino acids. In fact, it has been claimed that Cary's mammary AV difference studies on lactating cows in 1922 provided the first quantitative evidence of the importance of amino acids in protein synthesis in any biological system. When, in the 1950s, Crick and Watson showed how nuclear DNA directs the order of amino acid polymerization, a satisfactory scheme for explaining protein synthesis in all tissues seemed to have been achieved.

Protein synthesis is central to metabolism because all enzymes, which through their catalytic activity regulate cellular reactions, are proteins. At the time of parturition the stimulation of genes in mammary cells leads to the synthesis of the enzymes which direct lactose, fat and protein synthesis. So in mammary cells genes are directly instrumental in determining synthesis both of enzymes and of the specific milk proteins.

It has been noted already that in describing biological processes it is often necessary to draw arbitrary limits around particular aspects to facilitate discussion.

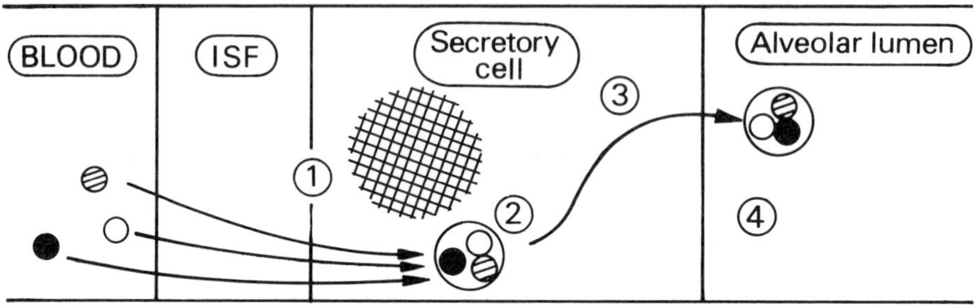

Figure 5.1. A generalized scheme of milk synthesis and secretion. (For explanation, see text.)

Figure 5.1 represents a generalized scheme of milk synthesis and secretion, from which four phases may be identified: (i) absorption of precursor molecules from the blood stream into the secretory cell; (ii) intracellular synthesis of milk components; (iii) translocation of these substances through the cell to the apical membrane, and (iv) secretion of the substances across the apical membrane into the alveolar lumen.

The first phase has been discussed in outline in Chapter 4: here we shall concentrate on the identity of the milk precursors and how they are assembled into milk components, the second phase. The latter will entail some reference to intracellular translocation (the third phase), but that subject, and the fourth phase, are dealt with more fully in Chapter 6.

5.2. Identification of milk precursors

Identification of the blood precursors of milk constituents has relied heavily on two sorts of study: (a) quantification of the uptake of material from blood, by use of the AV difference method (see section 4.5), and (b) measurement of the extent of incorporation of radioactive label from precursor material into secreted milk components. In the former case, most studies have been performed on large animals, like goats, cows and sheep, because of the relative ease of sampling from the appropriate blood vessels (which lie close to the body surface) and of making the blood flow measurements necessary to quantify the results (see section 4.5.1).

Many isotope studies have also been carried out on large ruminants, but some have employed isolated mammary tissue from either such species or from laboratory animals like rats and guinea-pigs. One technique which has proved to be particularly useful is *perfusion of the isolated mammary gland*. In this method, glands are removed (under anaesthesia) from lactating animals, which are then killed before recovering from the anaesthetic. The glands are installed in an apparatus in which blood is passed through the cannulated blood vessels by

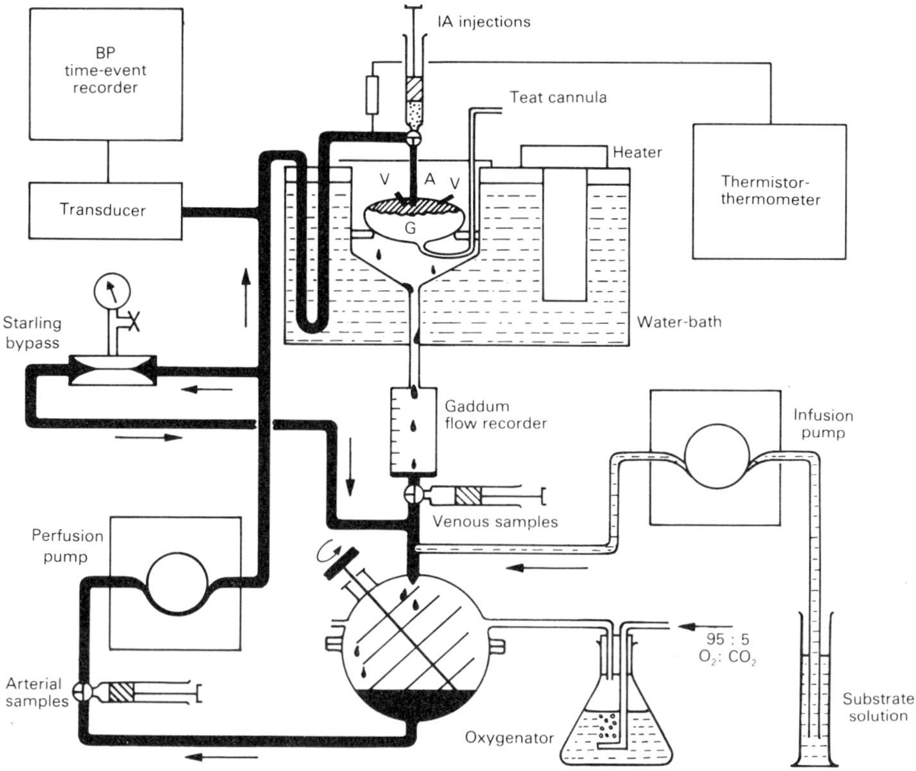

Figure 5.2. Diagram of circuit for perfusion of an isolated mammary gland of a guinea-pig. G, mammary gland; A, artery; V, veins.
(Source: Davis and Mepham, 1972.)

means of a pump (Figure 5.2). Adequate oxygenation of the blood and provision of substrates permit maintenance of the functional gland in isolation from many of the factors which normally control its activity, so that it is possible to study with ease the effects of, for example, alteration of blood flow rate, variation of substrate supply and effects of hormone administration. Such glands continue to secrete milk for several hours, although because they are denervated it is necessary to inject oxytocin to elicit the milk ejection response (see section 1.5). Many studies have also employed *in vitro* incubation of slices of mammary tissue, disaggregated cells or subcellular fractions. Comparison of the specific radioactivity of labelled precursors and reaction products in such preparations has enabled determination of the relative importance of different materials in the synthesis of milk constituents.

Reliable information on milk precursors requires both the uptake and isotopic ((a) and (b) above) approaches. It is, for example, possible for isotopic label to be incorporated into a milk constituent without there being any *net* synthesis (because some reactions might show complete reversibility at a constant rate), while *net* uptake (obtained from AV difference measurements) does not indicate the fate of absorbed material—it might be oxidized rather than incorporated into a milk component. Indeed, oxidation of an appreciable amount of material absorbed from blood is a process of obvious importance in the generation of ATP molecules to drive the cells' anabolic reactions.

Having discussed general principles involved in assessing substrate utilization by the mammary gland, we can now turn to a consideration of the specific precursors of the different milk constituents and the biosynthetic reactions involved.

5.2.1. Differences between ruminants and non-ruminants

It is, however, useful to consider at the outset a species-distinction which has important consequences for mammary metabolism in those animals which have been studied most extensively: the low glucose concentration and high acetate concentrations in blood of ruminant species by comparison with those in non-ruminants. Because of the activity of the microbial population of the rumen (comprising bacteria, protozoa and fungi), dietary components, like carbohydrates and proteins, are largely fermented to acetate, butyrate and propionate. Acetate is thus present at about 1.5 mM in cow blood, compared with 0.25 mM in rat blood, whereas the corresponding values for glucose are about 2.5 mM and 4.5 mM, respectively. Most of the glucose in ruminant blood is derived from propionate in the liver. Butyrate is largely converted to 3-hydroxybutyrate on absorption across the ruminal wall into blood, and the concentration of the latter in cow blood is over double that in rat blood. Amino acids in ruminant blood are derived from postruminal digestion of microbial protein, so that their concentrations are not so closely related to the dietary protein intake as they are in non-ruminants.

The distinction between cows and rats has been emphasized, but, in fact, these represent the opposite ends of a continuous spectrum. Thus, in non-ruminant herbivores, like rabbits and guinea-pigs, blood acetate concentrations are higher than in rats because of microbial activity in the caecum; the effects on their metabolism are consequently intermediate between the two extremes.

5.3. Protein synthesis

5.3.1. Amino acid metabolism

AV difference studies coupled with MBF measurements in both ruminant and non-ruminant species show that amino acids fall into two major categories, those for which the mammary uptake from blood plasma is adequate to supply all the corresponding amino acid residues in the secreted milk-specific proteins (roughly corresponding to the traditional *essential amino acid* (EAA) category) and those for which uptake is inadequate for the purpose (corresponding to the *non-essential* (NEAA) category). Results of experiments involving AV difference measurements, isotopes and perfused gland preparations indicate that EAA may be further divided into those in which uptake from blood and output in milk protein are almost exactly balanced (group I) and those for which uptake significantly exceeds output in milk (group II). In the latter case the amino acids must be metabolized in the gland, and there is strong evidence that, via transamination reactions, they provide nitrogen for the intramammary synthesis of inadequately absorbed NEAA, while their carbon skeletons are oxidized to CO_2. Examples of typical NEAA and group I and II EAA are serine, methionine and valine, respectively. In the few studies in which appropriate measurements have been made, it has been calculated that the mammary uptakes of total amino acid nitrogen and carbon balance quite closely with the outputs of these elements in milk protein, so that, although intracellular synthesis and catabolism of amino acids is substantial, taken as a whole, amino acids are the major, if not sole, precursors of the specific milk proteins.

In recent years, however, several observations have suggested that the situation may be more complicated. For example, in early or very late lactation in ruminants, estimates of blood plasma AV differences often seem to be too low to account for the observed rates of milk protein secretion. Some evidence has suggested that amino acids may be derived from sources other than the plasma, for example, from the red blood cells and from plasma peptides. Another observation of some interest is that rates of protein synthesis greatly exceed those evident from secretion rates, implying that synthesized proteins suffer rapid proteolysis in the mammary gland. (This does not, of course, invalidate AV difference measurements, since they only indicate *net* changes).

Figures 5.3a and 5.3b summarize current ideas on the intracellular metabolism of amino acids in mammary cells from the standpoints of carbon and nitrogen interrelationships, respectively. It is clear that both glucose and acetate derived from the blood contribute carbon to amino acid synthesis (as indicated by isotope experiments), but this is probably not so in a *net* sense (because of the good carbon balance, noted above). The figures show that in some species, at least, urea is produced by the mammary cells and diffuses into both blood and milk. Figure 5.3b indicates that amino acids are also involved in polyamine synthesis (see section 9.6.2).

5.3.2. *Amino acid transport*

Our discussion of amino acid utilization has inevitably caused us to run ahead of ourselves, because most of the studies involved were based on comparing uptake from blood with output of the completed product, milk protein. We need to retrace our steps to examine how amino acids get into the cells in the first place. Transfer of molecules from blood to the ISF is discussed in section 4.7, so here we are concerned with how amino acids gain access to the cell interior from the ISF.

Whereas the basal lamina (see section 2.3.7) offers little resistance, the cell membrane, being predominantly lipid in composition, effectively prevents entry of the hydrophilic amino acids by simple diffusion. Transport across the basal membrane thus depends on special *carrier systems*. It is only very recently that any studies of amino acid transport systems in mammary tissue have been made.

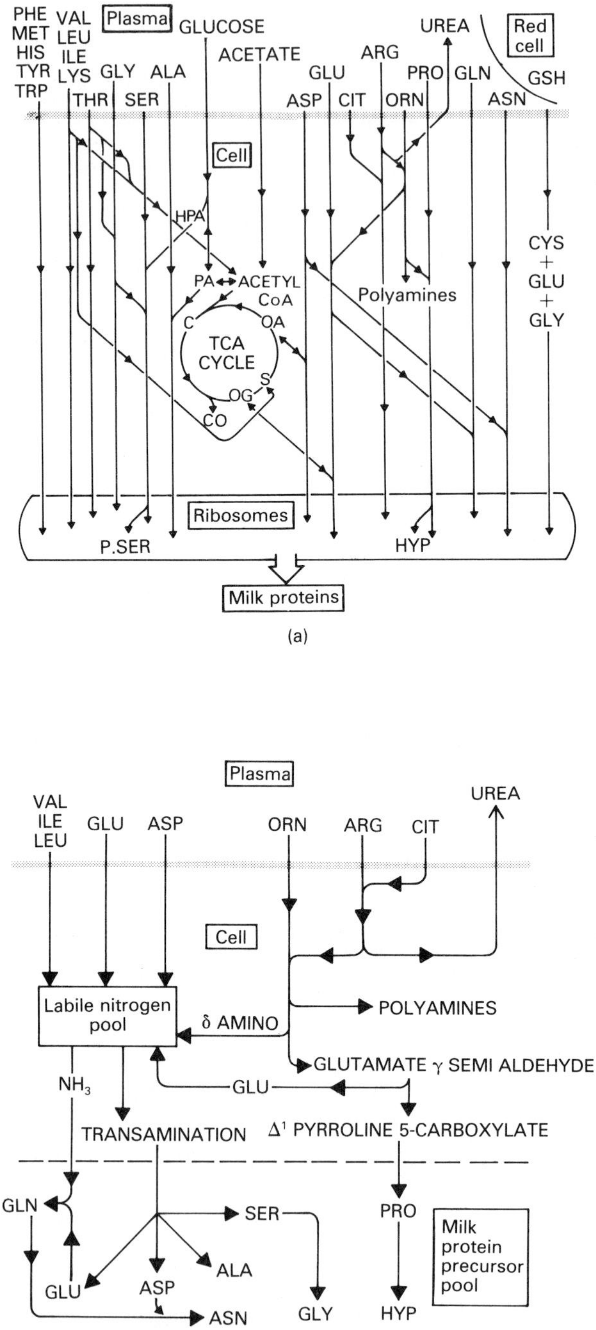

Figure 5.3. Summary diagrams of amino acid metabolism in mammary tissue. (*a*) Amino acid carbon interrelationships. (*b*) Amino acid nitrogen interrelationships.
(Source: Mepham, Gaye and Mercier, 1982.)

In other tissues several carriers have been described, which have variable affinity for different amino acids. For example, there is an *A system* (with a high affinity for alanine) and an *L system* (high leucine affinity). About six such systems have been characterized. It is envisaged that the carrier protein, located in the cell membrane binds with an amino acid in the ISF (in a manner akin to that involved between enzymes and substrates) and facilitates its transfer to the cell interior, where the amino acid is released leaving the carrier protein free to bind with another molecule. There is growing evidence for the presence of several such transport systems in mammary tissue. The fact that different amino acids enter cells on the same carrier protein implies that there is competition for entry. Thus, an amino acid at low ISF concentration which has a high affinity for a particular carrier may inhibit the cellular uptake of another amino acid of lower affinity, even if the ISF concentration of the latter is high. The factors controlling carrier protein synthesis and activity might thus exert considerable regulative control over milk synthesis as a whole.

5.3.3. *Milk protein synthesis*

In most milks the predominant milk proteins are the caseins. Since these are phosphoproteins, and, in some cases, glycoproteins, which aggregate into micelles (see section 3.3.1), description of their synthesis entails consideration of how the amino acids are combined and how the various non-protein components of the molecule are added. In theory, the synthesis of β-lactoglobulin and α-lactalbumin are somewhat less complicated, although some components of the latter are glycoproteins.

Milk protein synthesis is determined by the sequence of nucleotides in mRNA molecules transcribed from nuclear DNA. The details of the process in many tissues have been the subject of intense interest in recent years and theories as to the mechanisms operating, involving participation of rRNA and tRNA, will be found in any modern textbook of biochemistry. Quite recently, application to mammary tissue of the techniques of genetic engineering, by which, for example, short segments of nuclear DNA from one species may be spliced into the genes of another species (to form *recombinant DNA*), has added greatly to our understanding of the organization of the milk protein genome and the manner of its coordinated expression in lactation.

Expression of milk protein genes only occurs in association with lactation: the factors which trigger it are discussed in Chapter 9. Here, we need only note that the initiation and maintenance of gene expression appear to be crucially dependent on the hormonal milieu of the secretory cells.

5.3.3.1. Assembly of polypeptide chains This takes place on ribosomes. For those proteins destined for intracellular use (for example, most enzymes), the process occurs on ribosomes which are free in the cytosol, whereas in the case of 'export' proteins (that is, milk proteins) the ribosomes are attached to membranes of the endoplasmic reticulum (ER). The overall process of protein synthesis involves four major steps: (i) formation of aminoacyl-tRNA; (ii) initiation of the polypeptide chain; (iii) the elongation cycle, and (iv) chain termination.

At all stages, this *translation* process is controlled by information coded in RNA molecules.

5.3.3.2. Vectored transfer of secretory proteins Since milk proteins differ from other proteins in that they are destined to leave the cell, it may be reasoned that they must possess some feature by which they may be recognized by the appropriate translocating machinery. One feature likely to be important in this respect is the *signal sequence*, a peptide chain of 15–29 amino acids at the amino terminal end of the growing polypeptide chain. According to the *signal hypothesis*, the presence of this sequence causes the ribosome to bind to ER membrane, in which a

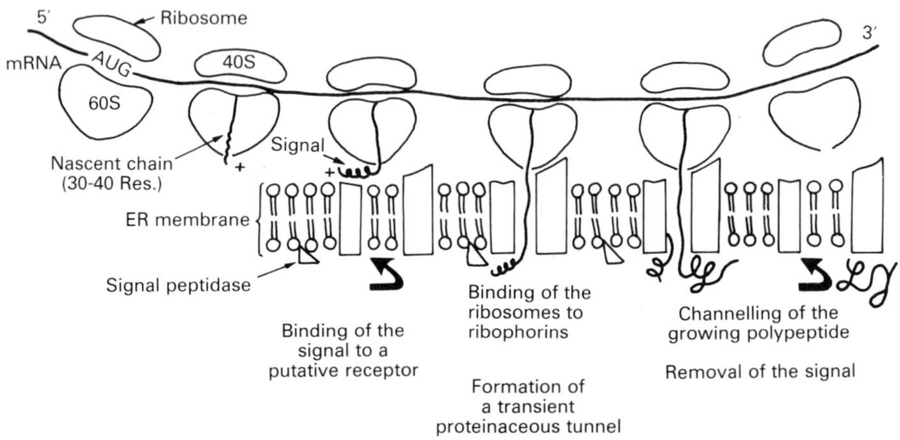

Figure 5.4. The signal hypothesis: a proposed mechanism for cotranslational crossing of the RER membrane by growing polypeptide chains.
(Source: Mercier and Gaye, 1983.)

'channel' forms, allowing the growing chain to enter the ER lumen (Figure 5.4). Subsequently, the signal sequence is split from the polypeptide chain by *signal peptidase*, an enzyme located on the luminal side of the ER membrane. Other vectorization theories have also been proposed.

5.3.3.3. Modifications of the polypeptide chain In addition to proteolytic processing (that is, removal of the signal peptide sequence), the polypeptide chain is subject to other covalent modifications: N- and O-glycosylation and O-phosphorylation. The route by which proteins cross the cell is discussed in Chapter 6, but here it is necessary to note that, after their synthesis and vectorization into the ER lumen, the proteins pass to the Golgi apparatus and thence, via secretory vesicles, to the apical membrane. Covalent modification must therefore occur at some point, or points, along this route. Such modifications may be either *co-translational* (occurring when chain elongation is in progress) or *post-translational*. As noted, proteolytic cleavage of the signal peptide is co-translational and this seems also to be the case for N-glycosylation, in which dolichol-linked oligosaccharides are enzymatically transferred to asparaginyl residues of the chain when these are present in the sequence code, Asn-X-Thr/Ser (where X is any amino acid except proline). The large oligosaccharide component may be 'trimmed' as it traverses the secretory pathway. Formation of disulphide bonds between adjacent sections of the chain, or between adjacent chains (as in κ-casein), may also be partly co-translational.

By contrast, O-glycosylation and O-phosphorylation appear to be post-translational events. Synthesis of the principal organ-specific glycoprotein, κ-casein, is believed to be effected by membrane-bound glycosyltransferases located in the Golgi apparatus, and three such enzymes have been described. O-phosphorylation involves transfer of the γ-phosphate of ATP to serine (or, less frequently, threonine) residues, occurring in the sequence, Ser/Thr-X-A (where X is any amino acid residue and A is an acidic residue, such as aspartic acid or a phosphorylated amino acid). Phosphorylation is effected by *casein kinases* which are chiefly located in the Golgi membranes, but the 'local environment' of the amino acid triplets is also important, a fact which may account for the lack of phosphorylation of β-lactoglobulin, in which the appropriate sites are made inaccessible by chain folding.

The Golgi complex is also the locus of casein micelle formation. In association with calcium, which is actively accumulated by Golgi vesicles, the polypeptide chains condense to form submicelles, and then micelles, prior to secretion. Micelles with a diameter of 140 nm may comprise 25 000 casein molecules.

5.3.4. *Non-secretory proteins*

As noted (in section 5.3.3.1), most proteins destined for intracellular use, such as enzymes, are synthesized on free ribosomes. Some, however, including lysosomal and integral membrane proteins, are synthesized on ribosomes bound to ER membrane. Like the secretory proteins, these are first synthesized complete with signal peptides, but subsequently they are diverted from the secretory path. In the case of lysosomal enzymes, diversion is effected by means of another signal, possibly a mannose-6-phosphate, which is added to the protein molecule before it reaches the Golgi complex. Lysosomal enzymes, enclosed within membrane-bounded lysosomes, are clearly important in intracellular digestion at involution, but they are also believed to be active in fully lactating glands to an extent which is hormone-dependent (see section 9.11).

5.4. Lactose synthesis

Of the major milk constituents, lactose has the simplest chemical structure (see section 3.5). Studies of glucose uptake by ruminant mammary glands in the early decades of the century indicated that more than enough glucose is absorbed from blood to provide all the carbon required for lactose synthesis (see section 4.5.1), but it was only when experiments with [U-^{14}C] glucose were performed in the 1950s that direct evidence was obtained that glucose is a precursor of lactose. For example, experiments with perfused mammary glands of goats showed that about 80% of lactose carbon is derived from blood glucose. Virtually all the glucose moiety of the molecule is formed from blood glucose, but the galactose moiety is also partly derived from acetate and glycerol carbon.

Glucose entry into the secretory cell from the ISF is thought to be effected by a specific carrier system in the basal membrane.

5.4.1. Enzymic steps in lactose synthesis

The sequence of reactions involved has proved to be more complicated, and indeed more interesting, than was originally envisaged. In the simplest scheme, where glucose is considered to be the sole precursor of both lactose moieties, glucose-l-phosphate arises by way of glycolytic pathway intermediates:

1. glucose + ATP $\xrightarrow[\textit{hexokinase}]{}$ glucose-6-P + ADP
2. glucose-6-P $\xrightarrow[\textit{phosphoglucomutase}]{}$ glucose-1-P

Alternatively, glucose-1-P may arise from carbon of, for example, acetate, by a longer route, though not in a *net* sense (see section 5.2). Glucose-1-P is subsequently converted to UDP galactose via two enzyme-catalysed steps:

3. glucose-1-P + UTP $\xrightarrow[\textit{UDP-glucosephosphorylase}]{}$ UDP-glucose + PPi
4. UDP-glucose $\xrightarrow[\textit{UDP-galactose 4-epimerase}]{}$ UDP-galactose

The terminal step in the synthesis of lactose was a subject of controversy for some time, that is, over whether glucose or glucose-1-P was the galactosyl acceptor. It has now been established that the former is involved. Thus the final reaction, catalysed by *lactose synthetase*, is:

5. UDP galactose + glucose → lactose + UDP

5.4.2. Lactose synthetase

This enzyme appears to be quite unique, both from the viewpoint of enzymology as a whole and because its specific properties result in its playing a key role in the control of milk secretion.

It was shown in 1966, from work with a soluble form of the enzyme present in milk, that for lactose synthesis to occur at physiological concentrations of glucose two components are required:

(a) the enzyme, *galactosyltransferase*, which is involved in glycoprotein synthesis (where it transfers the galactosyl moiety of UDP-galactose to a number of acceptors, such as oligosaccharides), but for which glucose itself is a very poor substrate, and

(b) the whey protein, α-lactalbumin, which, by associating with galactosyltransferase in a complex, increases its affinity for glucose.

Alpha-lactalbumin has thus been designated a *specifier protein*, by virtue of its property of changing the enzyme specificity. It achieves this by, effectively, creating a binding site specific for β-glucose, thus reducing the apparent K_m from molar to millimolar concentrations.

Although most studies have been performed on the soluble enzyme, in mammary cells galactosyltransferase is firmly bound to the inner surface of Golgi membranes, as would be expected in view of its role in glycoprotein synthesis. However, during lactation, α-lactalbumin, along with the other secretory proteins, is transported through the Golgi complex *en route* for the apical membrane. It is thus able to form a transient association with the membrane-bound galactosyltransferase and effect the change in its specificity required for lactose synthesis.

Lactose synthetase is only activated in the presence of bivalent cations. In the bovine enzyme there are two cationic binding sites: site I, which has highest affinity for Mn^{2+}, and site II which is less specific; the physiologically active controls for site II are a matter of uncertainty. Kinetic analyses suggest that the substrates and cofactors combine sequentially to form a *central complex*:

galactosyltransferase-Mn^{2+}-X^{2+}-UDP galactose-glucose-α-lactalbumin

(where X^{2+} is a bivalent cation), which then dissociates to form lactose. Energetically, the overall reaction appears to be irreversible, a factor of clear advantage in accumulating lactose at high concentration from low precursor concentrations.

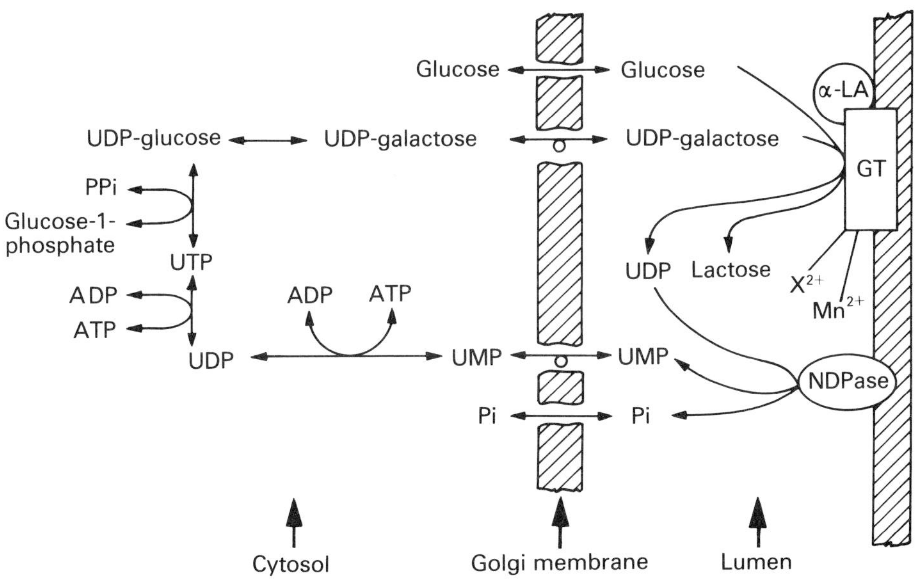

Figure 5.5. Proposed uridine nucleotide cycle supporting lactose synthesis. Double-headed arrows show reversible reactions or transmembrane movements via proposed pores (⟷) or carriers (⟷ with o). GT, galactosyltransferase; α LA, α-lactalbumin; NDPase, nucleoside diphosphatase; X^{2+} unknown cationic activator.
(Source: Kuhn, 1983.)

Figure 5.5 illustrates the topography of lactose-synthesizing enzymes. Evidence from several sources suggests that glucose enters the Golgi lumen via water-filled pores in the membrane, but the lactose molecule, because of its larger size, is unable to leave via the same route and thus accumulates in the lumen. While UDP and UDP-glucose do not appear to cross the membrane, it has been proposed that UMP and UDP-galactose do so via specific carrier systems. According to this scheme, a *uridine nucleotide cycle* can be envisaged (Figure 5.5), indicating the need for three ATP molecules to form one lactose from glucose. UDP is a powerful inhibitor of lactose synthetase, so that its conversion to UMP by membrane-bound nucleoside diphosphatase (NDPase), the distribution of which closely parallels that of lactose synthetase, seems to be an important element of the cycle.

Lactose is the major osmole of milk, and since, when in the cell, it is confined to the Golgi and secretory vesicle lumina, it osmotically draws water into them from the cytoplasm. In the majority of milks, water accounts for over 80% of milk volume, so that lactose synthesis exerts a crucial role in the secretion of milk as a whole (see section 6.3).

5.4.2.1. Evolutionary aspects Determination of the amino acid sequence of α-lactalbumin from several species reveals a close structural homology. Perhaps more interestingly, a close sequence homology also exists between α-lactalbumin and human and egg-white lysozymes, enzymes which hydrolyse bacterial cell wall mucopolysaccharides. It thus seems probable that α-lactalbumin and lysozyme are derived from a common ancestral gene.

Galactosyltransferase is widely distributed in both plant and animal tissues, in all cases showing low affinity for glucose. Nevertheless, galactosyltransferase derived not only from mammalian tissues, such as rat liver, but even that from onion roots, shows greatly increased glucose binding in the presence of α-lactalbumin, leading to lactose synthesis. It thus appears that, since selection pressures ensured the persistence of the role of galactosyltransferase in glycoprotein synthesis, evolution of lactose synthesis was entirely dependent on the appearance of α-lactalbumin.

5.5. Milk fat synthesis

The origin of animal fats was the subject of a protracted, and often rancorous, debate in the mid-nineteenth century. Dumas and Boussingault in France claimed that animals *could not* form fats from non-lipid material, whereas Liebig, in Germany, claimed that they *could*. Production of milk fat was central to this argument. When, finally, the ability of animals to synthesize fats was established beyond doubt, the much more restrained contributions to the debate of Lawes and Gilbert, in England, were seen to have provided crucial evidence. Nevertheless, the morphological site of milk fat synthesis and the nature of its precursors remained in doubt for many years. As with milk protein, it was only when sufficiently refined chemical procedures were devised that it became apparent the milk fats differed appreciably from both body fats and the lipids in the blood stream. Nevertheless, it was also apparent at this time that the chemical composition of milk fat is sensitive to a marked degree to the nature of the dietary fat.

Early AV difference studies in ruminants suggested that blood glycerides were absorbed by the udder in amounts adequate to account for all the milk fat secreted, but this was obviously not the whole story because milk fat was known to differ appreciably from blood lipids, particularly in respect of the high short-chain fatty acid content of the former (see section 3.4). In the early 1950s a theory was advanced which explained the fatty acid composition of ruminant milk fats as due to intramammary degradation of long-chain fatty acids by an oxidative process. But, concurrently, experiments were performed with radioactively labelled acetate which indicated that acetate carbon is incorporated into milk fatty acids with even-carbon numbers, up to a chain length of C_{16}.

5.5.1. *Milk fatty acid precursors*

We now know that the fatty acids of milk triacylglycerols (TG), which form the bulk of the fat in most species, are derived from two types of precursor:

fatty acids of blood lipids, the composition of which is markedly influenced by dietary fat intake, and
acetate and 3-hydroxybutyrate in ruminants or glucose (and, to some extent, depending on species, acetate, lactate and alanine) in non-ruminants.

Elucidation of the nature of the precursors and the synthetic pathways involved has stemmed from evidence acquired by several techniques, such as subcellular fractionation, enzyme kinetic studies and autoradiography.

5.5.1.1. Blood lipids AV difference and isotopic studies on lactating goats showed that the mammary gland extracts a large percentage (see section 4.5) of the fatty acids of blood plasma *chylomicrons* and *low density lipoproteins* (LDL). Since these exist in blood as large particles (Figure 5.6), it was a matter of some interest as to how this extraction is achieved. A clue was provided by the observation that the enzyme, *lipoprotein lipase* is present at higher concentration in mammary venous blood than in arterial blood, suggesting that it is released by the gland. The situation became clear with the report of EM histochemical studies on rat tissue which showed that chylomicrons become firmly embedded in mammary capillary endothelial walls, where they are hydrolysed by the lipoprotein lipase released at this site. Thus, fatty acids (invariably long chain, such as C_{16} and C_{18}) released by lipolysis, are absorbed across the basal membrane of the cells. Some *free fatty acids* (FFA) of plasma (which are bound to serum albumin) are also absorbed, equilibrating with those released by lipolysis, but under normal conditions there is no *net* AV difference of FFA across the mammary gland. It is now believed that in both ruminants and non-ruminants, milk fatty acids of chain length C_{18} and above are derived exclusively from blood lipids, while most of the C_{16} is also provided in this way. Some fatty acids are desaturated before incorporation into fat: for example, in goats, oleate is formed from stearate.

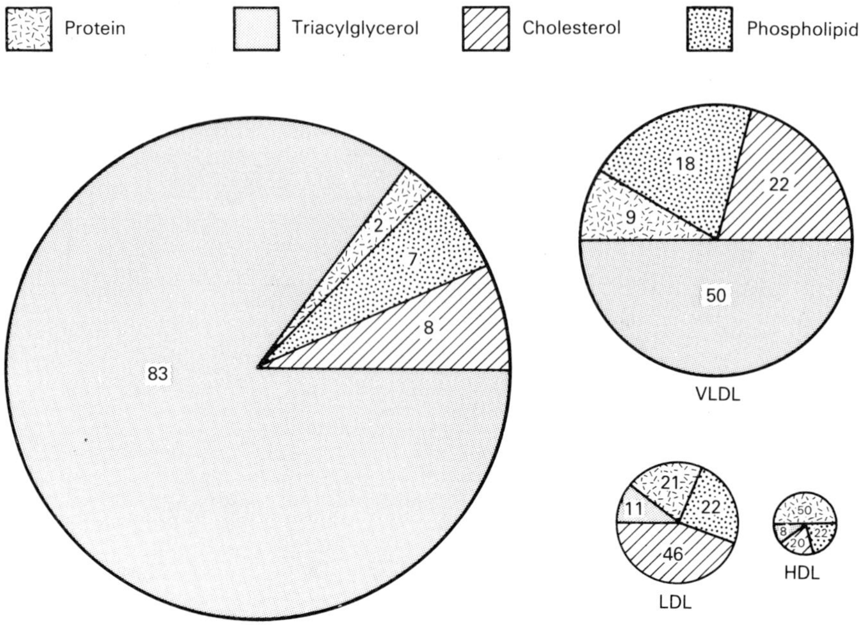

Figure 5.6. The composition of human serum lipoproteins. VLDL, very low density lipoproteins; LDL, low density lipoproteins; HLD, high density lipoproteins.
(Source: Gurr, 1983.)

The mechanisms by which FFA cross the basal membrane of the cell may involve diffusion and/or carrier-mediation.

5.5.1.2. Precursors of fatty acids synthesized in mammary cells In non-ruminant species the principal precursor of fatty acids up to C_{16} chain length is glucose. It is converted, via the glycolytic pathway, to pyruvate, which, following entry into mitochondria generates acetyl CoA. Acetyl CoA is the starting point of the *malonyl CoA pathway* by which fatty acids are synthesized, but it cannot be used directly because, while the fatty acid synthesizing enzymes are located in the cytosol, the mitochondrial membrane is impermeable to acetyl CoA. Instead, the acetyl groups cross the membrane, as it were, 'in disguise'. This is achieved by:

condensation of acetyl CoA and oxaloacetate within the mitochondria to form citrate
diffusion of citrate across the mitochondrial membrane into the cytosol, and
ATP-driven cleavage of citrate to acetyl CoA and oxaloacetate again by the enzyme *ATP citrate lyase* (Figure 5.7).

In ruminants the situation is quite different. In these animals ATP citrate lyase activity in the cytosol is very low, so that virtually no glucose carbon is used for fatty acid synthesis. (This appears to be an adaptation to low blood glucose concentrations, allowing the available glucose to be employed in other reactions,

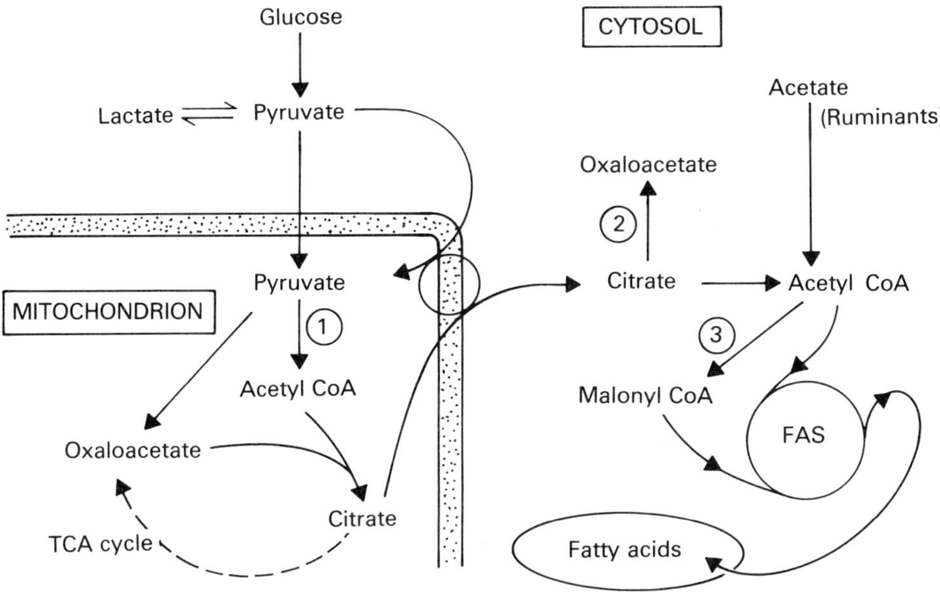

Figure 5.7. Sources of carbon for fatty acid synthesis in the mammary gland. Enzymes: 1., pyruvate dehydrogenase; 2., ATP citrate lyase; 3., acetyl CoA carboxylase; FAS, fatty acid synthetase.

such as lactose synthesis.) Instead, acetyl CoA derives from activation of acetate, which is present at high concentration in ruminants. Since in this case acetyl CoA is formed in the cytosol, it can be used directly for fatty acid synthesis (Figure 5.7). In non-ruminant herbivores (see section 5.2) fatty acid carbon is largely derived from glucose, but also, to some extent, from acetate. In ruminants, 3-hydroxybutyrate also acts as a minor precursor of milk fatty acids.

5.5.2. Fatty acid synthesis

We have seen that even-carbon numbered fatty acids of chain length C_{16} and below are synthesized from glucose and acetate carbon by an enzyme system located finally in the cytoplasm. Now we need to consider the reactions involved and the way in which the different fatty acid compositions, characteristic of the milks of different species, arise.

In fatty acid synthesis by the malonyl CoA pathway the first 'committed step' involves malonyl CoA formation from acetyl CoA by acetyl CoA carboxylase:

$$\mathrm{CH_3\text{-}CO\text{-}S.CoA + ATP + HCO_3^-} \xrightarrow[\text{Mg}^{2+}/\text{Mn}^{2+}]{\textit{acetyl CoA carboxylase}} \mathrm{^-OOC\text{-}CH_2\text{-}CO\text{-}S.CoA + ADP + Pi}$$

The second stage, elongation of acetyl CoA to long-chain fatty acids, by malonyl CoA, is catalysed by *fatty acid synthetase* (FAS), which is identical to that occurring in other tissues of the body. The overall reaction is:

$$\text{acetyl CoA} + 7\ \text{malonyl CoA} + 14\ (\text{NADPH} + \text{H}^+)$$
$$\downarrow$$
$$\text{palmitate} + 7\text{CO}_2 + 14\ \text{NADP}^+ + 8\text{CoA.SH} + 6\text{H}_2\text{O}$$

Thus, if the reaction goes to completion (which, as we shall see, is not always the case) palmitic acid is formed.

In addition to acetate, 3-hydroxybutyrate appears to contribute to this pathway, either by direct incorporation of C_4 units (butyryl CoA is a good 'primer') or following cleavage to C_2 units.

FAS is a large so-called polyfunctional protein (with a molecular weight of about 500 000), comprised of two identical polypeptide chains, arranged head to tail. Each chain possesses seven catalytic centres (Figure 5.8), and in the most widely accepted model a shuttle mechanism is suggested, whereby the growing acyl chain is transferred from one polypeptide chain to the other each time a two-carbon unit is added from malonyl CoA. This transfer is effected by the *acyl carrier protein* (ACP), which, though it possesses no enzymic activity itself, lies at the centre of the complex.

Following the purification of FAS from lactating mammary tissue in the early 1970s the question arose as to how the species-specific patterns of short-chain fatty acids are achieved, when the normal product of the reaction sequence is the C_{16} acid. Clearly, factors must be present which trigger premature chain-termination. By the late 1970s one such factor had been identified as *medium-chain acylthioesterase*, an enzyme which has been isolated from rabbit, mouse and rat mammary tissue. Addition of this enzyme to purified FAS changes the pattern of fatty acids synthesized from C_{16} to a mixture of C_8 and C_{10} acids. The situation in ruminants seems to be different in that fatty acids are released from FAS as acyl CoA esters instead of as unesterified acids, and even so, release probably only occurs in the presence of the microsomal triacylglycerol synthesizing system (see section 5.5.4). Thus, in ruminants, determination of chain-termination appears to reside in the FAS complex.

In addition to the malonyl CoA pathway, which accounts in large measure for synthesis of C_{10} to C_{16} fatty acids, other fatty acid synthetic pathways have been described. Prominent among these is that associated with ribosomes, which forms C_4 to C_8 acids from acetate and 3-hydroxybutyrate.

5.5.3. *Precursors of the glycerol moiety*

AV differences and isotopic experiments have shown that the glycerol backbone of the milk TG molecules is derived from three sources:

glucose, via glycolytic pathway intermediates
free glycerol of blood plasma
plasma lipids, following lipolysis at the capillary wall (see section 5.5.1.1).

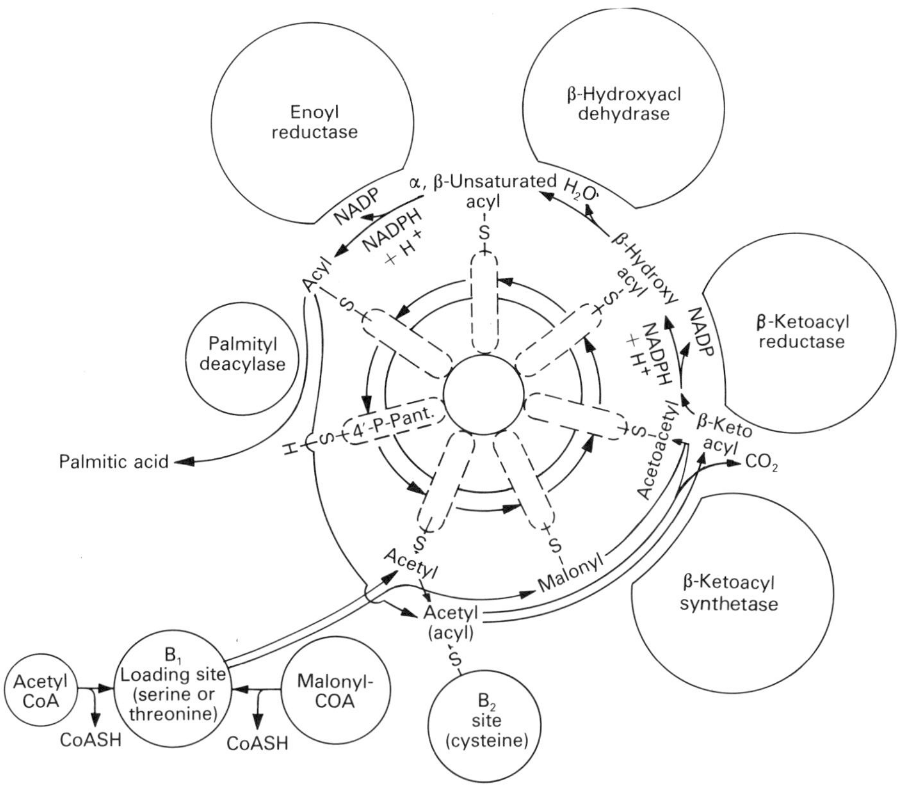

Figure 5.8. Organization of catalytic sites in fatty acid synthetase. The acyl carrier protein is at the centre of the complex.
(Source, Conn and Stumpf, 1976.)

There is evidence that in some cases lipolysis is incomplete, so that 2–monoacylglycerols may be absorbed by the cells and incorporated into milk fat.

5.5.4. *Triacylglycerol synthesis*

The final step in fat synthesis, esterification of glycerol to form TGs, most probably occurs on ER membranes. The process requires prior activation of the fatty acids to their acyl CoA thioesters by *ATP-dependent acyl-CoA synthetases,* but, as noted (see section 5.5.2), in ruminants the fatty acids are thought to be released from the FAS complex in this form.

The principal route of TG synthesis in mammary tissue is the *glycerol 3 phosphate pathway,* in which two molecules of acyl CoA ester are esterified to glycerol 3 phosphate to form phosphatidate, which, on the addition of a third acyl CoA molecule yields the triacylglycerol (Figure 5.9). The pathway has been demonstrated in both ruminant and non-ruminant tissues.

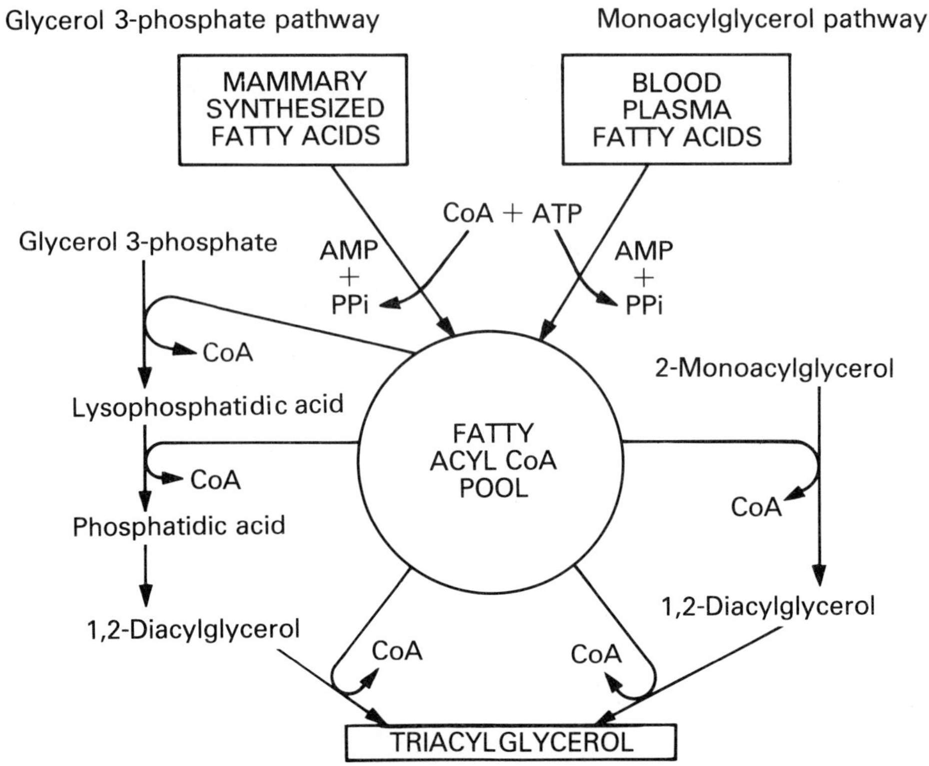

Figure 5.9. Pathways of milk triacylglycerol formation. Left: glycerol 3-phosphate pathway, right: monoacylglycerol pathway.

A minor route involves monoacylglycerol, derived from partial lipolysis of blood lipids. In this case, addition of one acyl CoA molecule forms 1,2 diacylglycerol, and of a second molecule, the completed TG (see Figure 5.9). The positioning of fatty acids on the glycerol backbone is not random: for example, shorter-chain fatty acids are located almost exclusively in the Sn_3 position.

Milk phospholipids and cholesterol form a very small percentage of the total milk fat, being largely confined to the MFGM. Whereas cholesterol may be partly derived from uptake of circulating lipoproteins (see Figure 5.6) and partly synthesized in mammary cells, milk phospholipids appear to be solely derived from intramammary synthesis. Figure 5.10 summarizes the precursors of milk TG.

5.6. Energy metabolism in secretory cells

All the anabolic processes discussed (polymerization of amino acids, esterification of glycerol; lactose, fatty acid and amino acid synthesis) require the provision

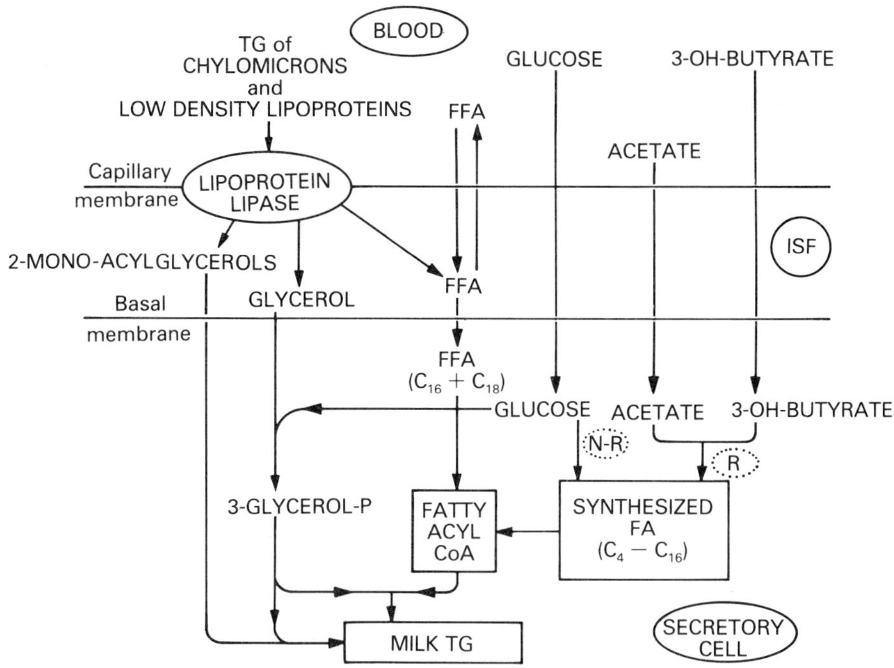

Figure 5.10. Summary of milk fat precursors. NR and R denote pathways existing in non-ruminants and ruminants, respectively. (For explanation, see text.)

of energy. This requirement is usually met by removal of the terminal phosphate or pyrophosphate group of ATP, but other nucleoside phosphates are also involved in particular cases, for example, UTP in lactose synthesis, GTP in protein synthesis and CTP in phospholipid synthesis. Since these nucleotides acquire phosphate from ATP, we are here principally concerned with pathways of ATP synthesis.

Mammary cells produce ATP by the same routes as those described for most animal cells: glycolysis, the pentose phosphate pathway and the tricarboxylic acid (TCA) cycle. Thus, our main interest is with the relative importance of the different pathways in mammary tissue and with the types of substrate catabolized in them.

It was apparent from section 5.2.1 that ruminants and non-herbivorous non-ruminants represent two ends of a spectrum in respect of blood concentrations of glucose (low in ruminants) and acetate (high in ruminants), and that this has consequences for the nature of the precursors of milk fatty acids in these species. It is hardly surprising that such differences also affect the type of fuel oxidized by cells of the different species. Furthermore, fatty acid synthesis depends on a supply of reducing equivalents in the form of NADPH (see section 5.6.2): corresponding species differences also exist in *their* routes of supply.

5.6.1. *ATP provision*

In all species studied the mammary uptake of glucose exceeds its requirement for the synthesis of lactose and for the glycerol of milk TG. Isotope studies show that there is an appreciable transfer of ^{14}C activity from glucose to CO_2, so that there is little doubt that glucose is an important metabolic fuel for mammary cells. There are, however, predictable species differences in the quantitative importance of glucose. Thus, studies on cows showed that about 25% of the CO_2 produced by the mammary glands was derived from glucose oxidation, whereas in sows the figure was about 55%. The apportionment of CO_2 production to particular metabolic pathways is a task fraught with technical difficulties, but a common feature emerging from several studies is the prominence of the pentose phosphate pathway as a pathway of glucose metabolism in both ruminant and non-ruminant species. Glycolysis seems to be only a minor pathway in ruminant glands but studies on non-ruminants suggest that glycolysis and the pentose phosphate pathway are equally active.

Estimates of glucose oxidation in ruminants, even allowing for the inherent difficulties associated with interpreting isotopic data, suggest that the oxidative role of the TCA cycle is only of minor importance. Consistent with that conclusion is the enzymological evidence which suggests that triose phosphate produced by the pentose phosphate pathway (which in non-ruminants is further metabolized via pyruvate and the TCA cycle) is recycled via fructose bisphosphate to glucose-6-phosphate, thereby increasing the catabolic efficiency of the pentose phosphate pathway (Figure 5.11).

Acetate accounted for about 30% of the CO_2 produced by cows' udders in one study, whereas the corresponding figure for sows was 2%: in non-ruminant herbivores, like guinea-pigs, the figure lies between the two extremes. Although glucose and acetate contribute roughly equally to mammary CO_2 production in ruminants, acetate is probably more important in generating ATP because its oxidation, proceeding via the TCA, is linked to oxidative phosphorylation, whereas (as noted above) glucose carbon may be largely recycled through the pentose phosphate pathway.

Amino acids may contribute carbon for oxidation via intermediates of the glycolytic and TCA pathways, as evidenced by $^{14}CO_2$ production in experiments in which ^{14}C amino acids have been administered to mammary tissue preparations from several species. But, as noted above (see section 5.3.1), there is usually a good balance between total amino acid carbon uptake and the output of milk protein carbon, so that it is questionable whether amino acids are a significant source of mammary CO_2 in a *net* sense.

Oxidation of fatty acids, 3-hydroxybutyrate and lactate appears to be minimal in fed ruminants, although it may become significant during starvation.

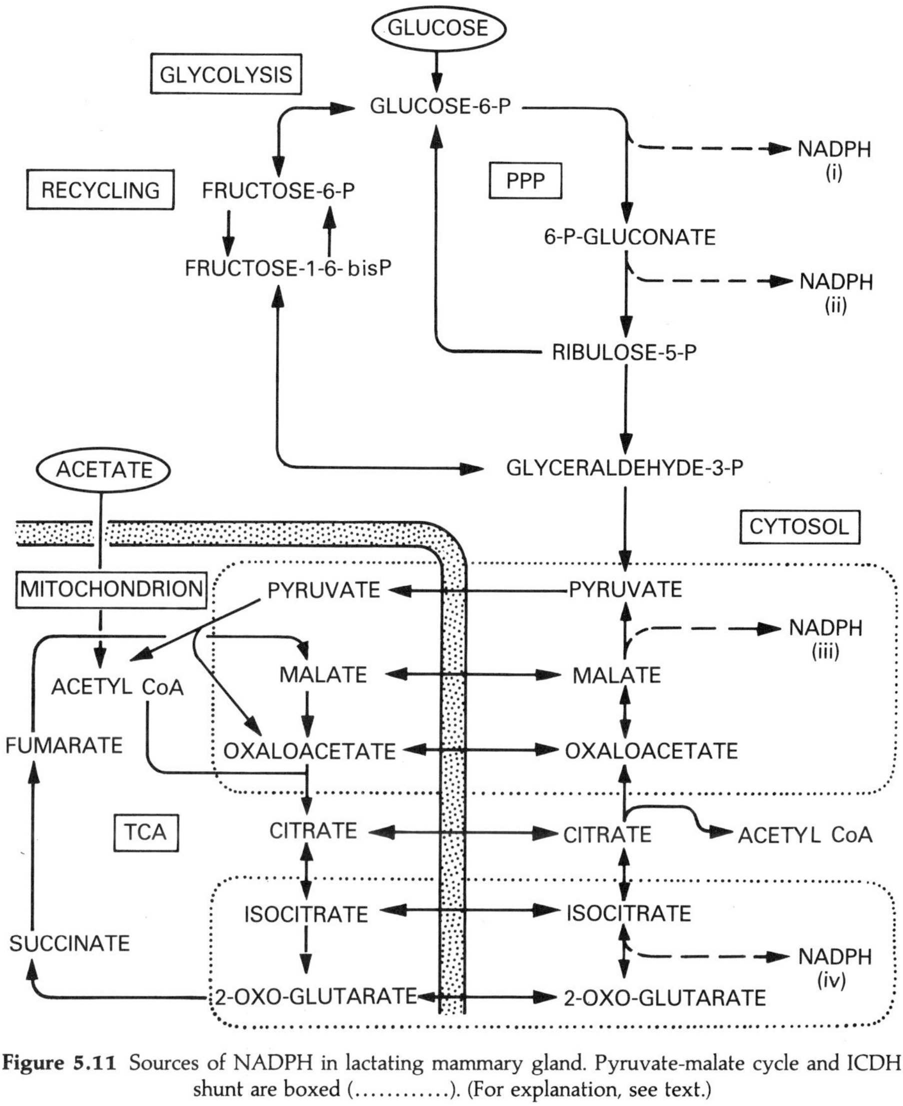

Figure 5.11 Sources of NADPH in lactating mammary gland. Pyruvate-malate cycle and ICDH shunt are boxed (…………). (For explanation, see text.)

5.6.2. NADPH provision

Since fatty acid synthesis takes place in the cytosol, the required NADPH molecules must be made available in this cell compartment. Four cytosolic enzymes have been identified which furnish NADPH:

glucose-6-phosphate dehydrogenase
6-phosphogluconate dehydrogenase
NAD-malate dehydrogenase, and
NADP-isocitrate dehydrogenase.

The available information suggests that their relative importance in mammary tissue is highly species-variable.

The first and second enzymes above are both active in the pentose phosphate pathway, in which three molecules of glucose-6-phosphate are oxidized to ribulose-5-phosphate, generating six molecules of NADPH. The three molecules of ribulose-5-phosphate are then rearranged to regenerate two molecules of glucose-6-phosphate and one molecule of triosephosphate (that is, glyceraldehyde-3-phosphate). Since there is a *net* consumption of only one molecule of glucose-6-phosphate, the outcome of the reaction sequence may be summarized as:

$$\text{glucose-6-phosphate} + 6\ \text{NADP} \rightarrow 6\ \text{NADPH} + 3\text{CO}_2 + \text{triose P}$$

The pentose phosphate pathway is active in mammary tissue of both ruminants and non-ruminants, and represents a major source of supply of reducing equivalents in the form of NADPH.

In non-ruminants, triosephosphate is further metabolized by the TCA cycle, but in ruminants the postulated recycling of fructose-6-phosphate (see section 5.6.1 and Figure 5.11) would result in increased efficiency of NADPH production. Thus, with recycling, oxidation of glucose-6-phosphate is complete and the NADPH yield is doubled:

$$\text{glucose-6-phosphate} + 12\ \text{NADP} \rightarrow 6\text{CO}_2 + 12\text{NADPH}$$

The advantage to ruminants of recycling is obvious: a given rate of NADPH production can be sustained with only half the glucose utilization rate required in its absence.

NADPH-malate dehydrogenase (the third enzyme in our list) exhibits significant activity in mammary tissue of rats and mice. Its utility in NADPH production involves its participation in the so-called *pyruvate-malate cycle*, in which reducing equivalents are transported across the mitochondrial membrane (see Figure 5.11). Current evidence suggests that the cycle only operates in these two species: it is not active in ruminants.

Ruminant mammary tissue can synthesize fat in the absence of glucose. This fact points to the existence of another source of cytosolic NADPH, in addition to the pentose phosphate pathway. While current evidence is rather indirect, a system involving NADP-isocitrate dehydrogenase (ICDH, for short), the fourth enzyme in our list, seems the most likely candidate for this role. The activity of the enzyme is some 20-fold greater in cow and sheep tissue than in that of rats, and the claim for its importance in ruminants is consistent with the fact that citrate is readily available in the cytosol because of the absence of ATP citrate lyase (see section 5.5.1.2). As currently conceived the so-called *ICDH-shuttle* allows mitochondrial NADH to be transhydrogenated with cytosolic NADP, isocitrate acting as a carrier of reducing equivalents between the two cell compartments (see Figure 5.11). Evidence from studies on cows and goats suggests that the ICDH shuttle may be capable of providing half of the NADPH required (the rest probably being generated by the pentose phosphate pathway).

Further reading

Annison, E. F. (1983). Metabolite utilization by the ruminant mammary gland, in *Biochemistry of Lactation*, pp. 399–436, ed. Mepham, T. B. (Amsterdam and New York: Elsevier).

Bauman, D. E. and Davis, C. L. (1974). Biosynthesis of milk fat, in *Lactation*, Vol. 2, pp. 31–75, eds Larson, B. L. and Smith, V. R. (New York and London: Academic Press.)

Craig, R. K. and Campbell, P. N. (1978). Molecular aspects of milk protein biosynthesis, in *Lactation*, Vol 4, pp. 387–404, ed. Larson, B. L. (New York and London: Academic Press.)

Davis, C. L. and Bauman, D. E. (1974). General metabolism associated with the synthesis of milk, in *Lactation*, Vol. 2, pp. 3–38, eds Larson, B. L. and Smith, V. R. (New York and London: Academic Press.)

Dils, R. R. (1983). Milk fat synthesis, in *Biochemistry of Lactation*, pp. 141–57, ed. Mepham, T. B. (Amsterdam and New York: Elsevier.)

Dils, R., Clark, S. and Knudsen, J. (1977). Comparative aspects of milk fat synthesis. *Symp. Zool. Soc. London*. **41**, 43–55.

Dils, R. and Parker, D. S. (1982). Metabolic aspects of lactation and the supply of nutrients to the young, in *Biochemical Development of the Fetus and Neonate*, pp. 571–88, ed. Jones, C. T. (Amsterdam: Elsevier.)

Ebner, K. E. and Schanbacher, F. L. (1974). Biochemistry of lactose and related carbohydrates, in *Lactation*, Vol. 2, pp. 77–113, eds Larson, B. L. and Smith, V. R. (New York and London: Academic Press.)

Fleet, I. R. and Mepham, T. B. (1983). Physiological methods used in the study of mammary substrate utilization in ruminants, in *Biochemistry of Lactation*, pp. 469–91, ed. Mepham, T. B. (Amsterdam and New York: Elsevier.)

Hawke, J. C. and Taylor, M. W. (1983). Influence of nutritional factors on the yield, composition and physical properties of milk fat, in *Developments in Dairy Chemistry*, Vol. 2, pp. 37–81, ed. Fox, P. F. (London and New York: Applied Science Publishers.)

Jones, E. A. (1978) Lactose biosynthesis, in *Lactation*, Vol. 4 pp. 371–85, ed. Larson, B. L. (New York and London: Academic Press.)

Kuhn, N. J. (1983). The biosynthesis of lactose, in *Biochemistry of Lactation*, pp. 159–76, ed. Mepham, T. B. (Amsterdam and New York: Elsevier.)

Mepham, T. B., Gaye, P. and Mercier, J. C. (1982). Biosynthesis of milk proteins, in *Developments in Dairy Chemistry*, Vol. 1, pp. 115–56, ed. Fox, P. F. (London and New York: Applied Science Publishers.)

Mercier, J. C. and Gaye, P. (1983). Milk protein synthesis, in *Biochemistry of Lactation*, pp. 177–227, ed. Mepham, T. B. (Amsterdam and New York: Elsevier.)

Patton, S. and Jensen, R. G. (1976). *Biomedical Aspects of Lactation* (Oxford and New York: Pergamon.)

Smith, G. H., Crabtree, B. and Smith, R. A. (1983). Energy metabolism in the mammary gland, in *Biochemistry of Lactation*. pp. 121–40, ed. Mepham, T. B. (Amsterdam and New York: Elsevier.)

6 Secretory Processes

6.1. Introduction

We have seen that milk is composed of a large number of chemicals, some of which are synthesized in the secretory cells whereas the remainder pass from blood to alveolar lumina without chemical change. The purpose of this chapter is to examine the ways in which the different milk constituents traverse the mammary epithelium.

Evidence from light microscopy in the nineteenth century gave rise to a succession of theories of milk secretion. For example, Virchow, the founder of 'cellular pathology', thought that mammary cells suffered complete degeneration (*holocrine* secretion) so that milk consisted largely of dispersed cellular material: continuing secretion thus depended on continuing cell division. Improved microscopical techniques then led to the view that only the apex of the cell was lost (*apocrine* secretion), the bulk of the cell remaining intact and subsequently reaccumulating secretory material at the apical pole until the next phase of secretion. Towards the end of the century a *merocrine* theory was proposed, in which cell membranes remain intact and no loss of cellular integrity is involved.

It was only when the electron microscope was employed in such studies, in the 1950s, that it became possible to resolve the conflict of ideas. It should not be imagined, however, that there is unanimity among electron microscopists concerning the mechanisms of milk secretion; as we shall see, there is still much debate. Our current understanding of secretion is not, in any case, based solely on EM evidence. The availability of radioactive isotopes in the last 30 years has permitted a wide range of studies. For example, in the technique of autoradiography, the time-course of passage of material through cells may be assessed by locating intracellular sites of the isotope in mammary tissue removed from laboratory animals which have been killed at serial times after administration of a radioactive milk precursor. Moreover, in dairy ruminants, the different time-courses of isotope appearance in different milk constituents, following intravenous injection of labelled precursors, has suggested that three principal secretory mechanisms occur, each with a distinctive transfer rate. Two other transfer processes are involved to a minor degree in established lactation, but their importance is greater in the colostrum-forming and involuting gland, and in certain disease states. Whatever the process involved, transepithelial transfer must be either transcellular or paracellular (see section 4.9.3).

6.2. Milk-specific proteins

The earlier discussion of milk protein synthesis (see section 5.3) proceeded to the point where, following vectorization into the ER lumen, the polypeptide chains are transferred to Golgi lumina. The route of transfer from ER to Golgi has not been defined with certainty. It is possible that lumina of the ER and Golgi apparatus are connected or, alternatively, that small vesicles bud from the ER and subsequently fuse with the Golgi membranes. In either case, casein molecules aggregate in the Golgi cisternal lumina in the form of micelles.

Lumina at the nuclear face of the Golgi apparatus (Figure 6.1) are termed *cis* cisternae; those at the apical face *trans* cisternae. Proteins appear to enter the complex at the *cis* face and progress, undergoing post-translational modification,

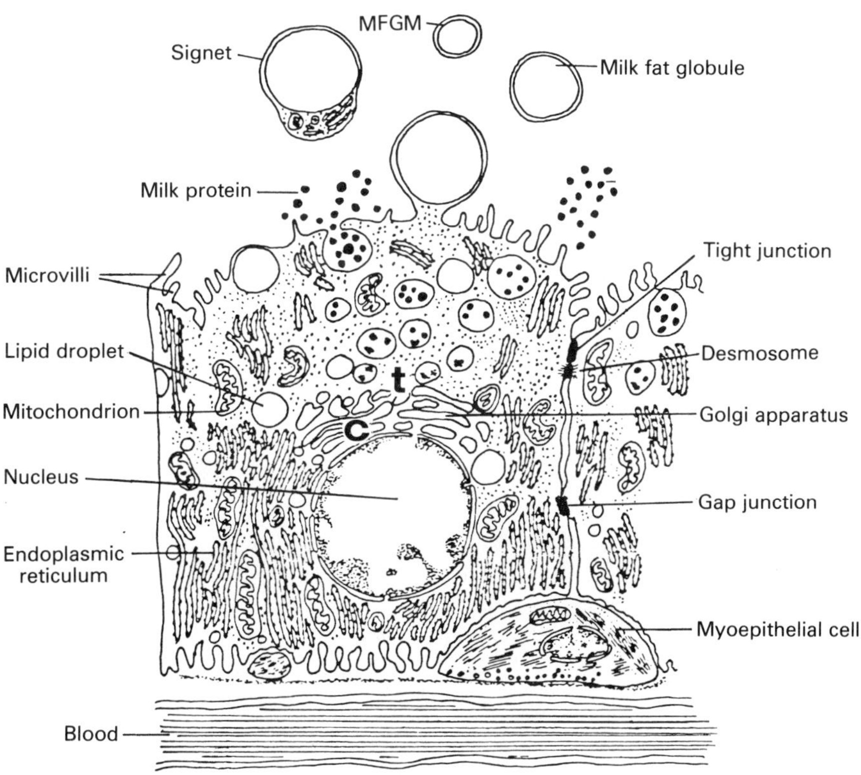

Figure 6.1. Schematic representation of a mammary secretory cell as interpreted from electron micrographs. c, *cis* face of Golgi apparatus; t, *trans* face.
(Adapted from Linzell and Peaker, 1971).

towards the *trans* face. Transfer between adjacent Golgi cisternae is thought to be achieved by budding and subsequent fusion of vesicles: this process is discussed more fully below (see section 6.8).

In the apical cytosol there are numerous protein-containing secretory vesicles (see Figure 6.1). EM studies suggest that they move to the apical plasmalemma and fuse with it releasing their contents by *exocytosis*, formerly called *reverse pinocytosis* (Figure 6.2). Two questions immediately arise: how is intracellular translocation achieved? and what determines the fusion of secretory vesicle and

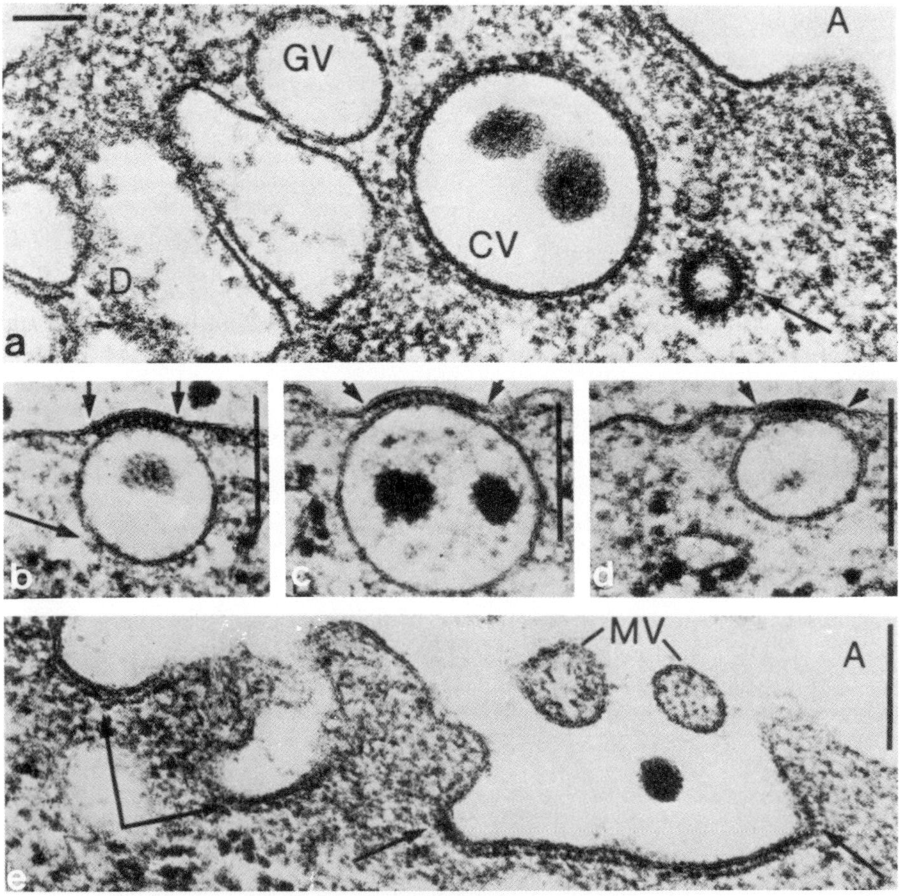

Figure 6.2. Electron micrographs of the apical region of mammary secretory cells of a rat, showing processes involved in secretion of casein micelles. (*a*) The major types of vesicle are identified: GV, Golgi-apparatus associated vesicles with a smooth surface; CV, coated vesicles containing casein; arrow, smaller vesicles also coated on cytoplasmic face. (× 97 750). (*b*), (*c*), (*d*) Pre-exocytotic attachment plaques forming between secretory vesicles and apical plasma membrane (short arrows): coat material is still present on some vesicles (long arrow in (*b*)). Magnifications: (*b*) × 85 000; (*c*) × 89 250; (*d*) × 93 500. (*e*) integration of coat material of secretory-vesicle membrane into apical plasma membrane (arrows); microvilli (MV) appear in cross section. × 95 000. (From Franke, 1976; (reproduced from *The Journal of Cell Biology*, 1976, **69**, 173–195, by copyright permission of the Rockefeller University Press and W. W. Franke.)

apical membranes? Current ideas on intracellular transport of vesicles invoke participation of cytoskeletal elements—microtubules and microfilaments (see section 2.3.4). In mammary cells these structures show an apicobasal orientation, suggesting that they may act as 'guides' for vesicular movement. Administration of antimicrotubular agents, like *colchicine,* inhibits milk secretion, but recent evidence indicates that caution is required in interpreting this observation because lumicolchicine, a chemical very similar to colchicine but lacking its anti-microtubular property, also inhibits secretion, so that inhibition does not necessarily occur at the microtubular level. Alternative theories of vesicle transport implicate a process of simple physical displacement as new vesicles bud from the Golgi complex or an 'electrophoretic' process, dependent on a transcellular potential gradient. Good evidence for any theory has not yet been obtained.

However they arrive there, secretory vesicles would seem to become attached to the cytoplasmic face of the apical plasmalemma. The vesicles have a distinctive coat on their outer surface (Figure 6.2a) which appears to react with the appropriate receptors on the apical membrane (*pre-exocytotic attachment plaques*) forming a series of regularly-spaced bridges (Figure 6.2b–d). Presumably these bridges, and the contiguous vesicle and apical membrane material, are subsequently eliminated and the vesicular contents released, but the process seems to be very rapid and it has proved difficult to visualize the details of the sequence by EM. But it is clear that secretory vesicle membrane becomes incorporated, however briefly, into the apical membrane as a consequence of exocytosis (Figure 6.2e).

Latterly, however, an alternative mechanism of protein secretion has been proposed in which secretory vesicles become linked in a 'ball and socket' manner and assume a linear orientation (Figure 6.3). According to this scheme, protein granules are transported through the lumina of a contiguous sequence of vesicles, so that only the most apical vesicle fuses with the apical membrane. The process has been called *compound exocytosis.*

Thus, if we put together the events of milk protein synthesis (see section 5.3.3) and secretion, the whole can be said to consist of eight steps: transcription, translation, segregation, modification, concentration, packaging, storage and exocytosis. These are summarized schematically in Figure 6.4.

6.3. Lactose and water secretion

Since lactose is synthesized in the Golgi lumina (see section 5.4.2), and is too large a molecule to escape into the cytosol, it would appear to share the same secretory route as the specific milk proteins. Osmotic entry of water leads to swelling of the Golgi and secretory vesicles, although there is some doubt as to whether further osmotic equilibration also occurs across the apical membrane.

6.4. Calcium, phosphate and citrate secretion

These ions are largely associated with the casein micelles and are thus secreted by the 'Golgi complex–secretory vesicle' pathway.

Figure 6.3. Compound exocytosis: a proposed mechanism for release of secretory vesicle contents. (*a*) Vesicles assemble in a chain by a 'ball and socket' interaction. (*b*) Linked vesicles fuse, resulting in a single luminal space continuous with the alveolar lumen. (*c*) Emptying of the vesicular chain causes collapse and subsequent fragmentation of the membrane.
(Source: Keenan and Dylewski, 1985.)

Calcium is thought to be actively transported into the Golgi lumina by a calcium-sensitive ATPase: while cytosolic calcium concentrations are probably less than 1 μM, micelle formation requires its presence at millimolar concentrations. Phosphate, on the other hand, is generated in the Golgi lumina during lactose synthesis (see section 5.4.2). Citrate is synthesized in the secretory cell (see section 5.5.1.2) and is believed to be actively accumulated by the Golgi apparatus.

Thus, milk proteins, lactose, calcium, phosphate and citrate are all secreted via a common path. This notion finds support in the observation that frequently-milked goats, receiving intravascular infusions of radioactively labelled milk precursors, showed maximum specific radioactivities (SRAs) for all these milk constituents 2–3 h later. However, for Na^+, K^+ and Cl^- ions, maximum SRAs were achieved in milk within 1 h of injection, suggesting that a different transfer process is involved.

6.5. Sodium, potassium and chloride secretion

Our understanding of the processes involved in monovalent ion secretion is largely based on studies using isotopic tracers and potential difference (PD)

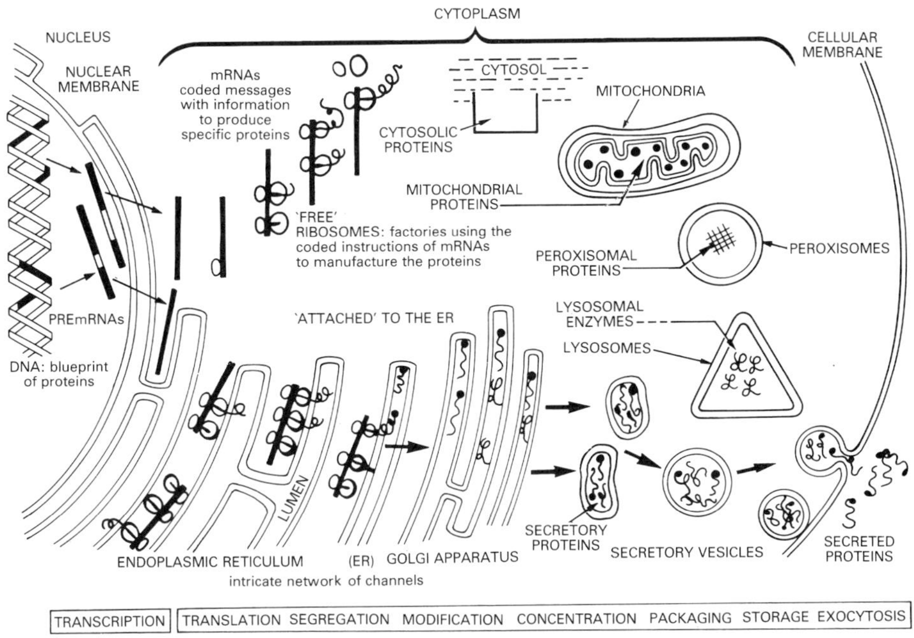

Figure 6.4. Schematic representation of the intracellular transport of proteins in mammary cells. (Source: Mercier and Gaye, 1982.)

measurements. Figure 6.5 shows that by comparison with ECF, milk contains less Na^+ and Cl^- and more K^+, with the milk concentrations of K^+ and Na^+ being in the ratio 3:1 for most species. Measurements of the intracellular (ICF) concentrations of these two ions show that, while they are higher than in milk, the 3:1 ratio also applies here. PD measurements indicate that the intracellular potential is markedly negative to milk, and, to an even greater degree, to ECF (Figure 6.6). In the scheme illustrated in Figure 6.7 such observations are consistent with K^+ and Na^+ transport occurring by a two-stage process: entry into the cell by means of a typical *Na^+–K^+-activated ATPase* on the basolateral cell membrane, followed by diffusion across the apical membrane in accordance with the prevailing electrochemical gradient. The presence of a Na^+–K^+-activated ATPase on the basolateral membranes and its absence on the apical membrane has been confirmed by histochemical studies of lactating mammary tissue. Since the Golgi-derived secretory vesicle membrane fuses with the apical membrane during exocytosis, it is surmised that the ions also diffuse into Golgi lumina, which may be regarded as, effectively, 'extracellular' spaces.

The concentrations of K^+ and Na^+ in milk are thus maintained at low levels by the apical membrane PD. But what generates this PD? A clue to the answer was provided by experiments with isolated perfused mammary glands (see section 5.2). When glands were deprived of glucose, not only was milk yield depressed,

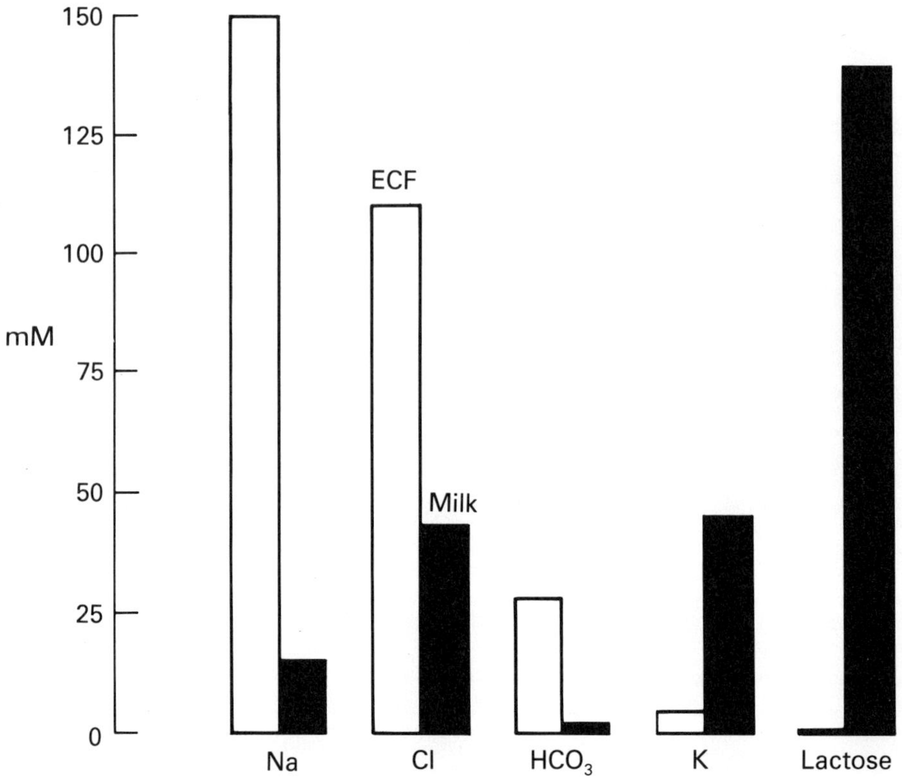

Figure 6.5. Concentrations of certain milk components in ECF and milk. (Source: Peaker, 1978).

as would be expected from the role of glucose as a precursor of the major osmole, lactose, but also the apical membrane PD decreased (see Figure 6.6: low milk [lactose]) and milk K^+ and Na^+ concentrations increased. These results imply that the PD is 'water-flow induced', that is, the osmotic flow of water across the Golgi and apical membranes, bearing fixed charges, causes charge-separation such that the luminal membrane becomes electrically positive. The K^+ and Na^+ ions are thus distributed across the membrane as predicted by classical streaming potential theory.

The secretion of chloride ions does not fit so neatly into this scheme. Since both the electrical and chemical gradients would tend to drive Cl^- into milk, it has been proposed that its low milk concentration is maintained by active reabsorption (Figure 6.7), although other explanations are conceivable.

6.5.1. A modification to the theory

While the above theory has received widespread acceptance, a report appearing at the time of writing has called it into question. Studies on mouse mammary tissue suggest that the K^+:Na^+ ratio is *not* the same in ICF and milk, and hence

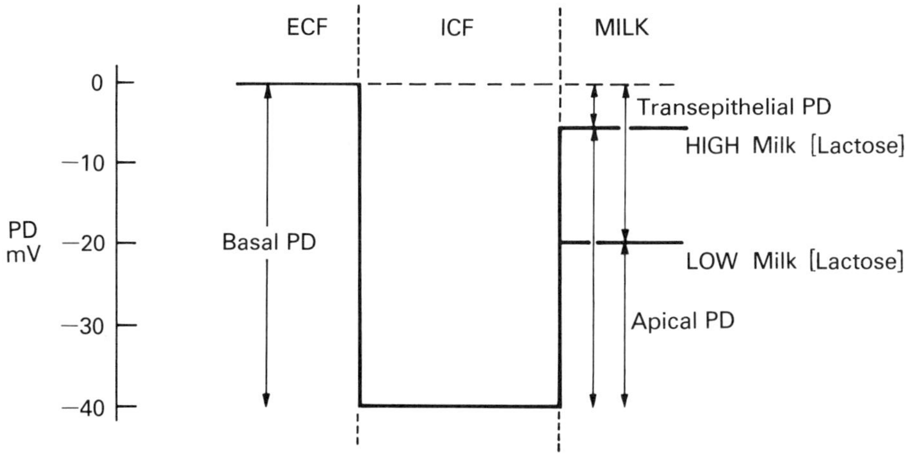

Figure 6.6. Electrical potential profile across the lactating goat mammary gland, showing the probable difference in apical PD between goats with a relatively high milk lactose concentration and those with low milk lactose.
(Source: Peaker, 1978.)

that an active transport system for Na^+ and/or K^+ exists at the apical membrane. Clearly, the matter is unresolved; and it again emphasizes the *provisional* nature of scientific theories.

6.6. The paracellular route

The ion movements described are the principal determinants of their milk concentrations in established lactation, since ductal reabsorption is negligible (see section 4.9.2). But in the parturient and involuting gland, and under conditions of excessive alveolar distension or mastitis, the ionic composition may be markedly different. In such cases, milk Na^+ and Cl^- concentrations rise and those of lactose and K^+ decline. The explanation is that the tight junctions cementing together adjacent cells become leaky and allow partial equilibration of milk and ECF via the paracellular route (see section 4.9.3).

The rabbit is unusual in that a paracellular route is present in this species throughout lactation, the atypically low K^+:Na^+ ratio declining to even lower values in later lactation.

6.7. Secretion of milk fat

Autoradiographic studies reveal that labelled fatty acids become associated with fat droplets in the secretory cell within minutes of their intravenous injection. This implies that TG molecules can be added to existing intracellular lipid droplets in the course of their transport to the apical membrane and that this process is very rapid. By contrast, the appearance of labelled fat in secreted milk reaches a maximum SRA 5–7 h after injection of fat precursors into goats, which is a much

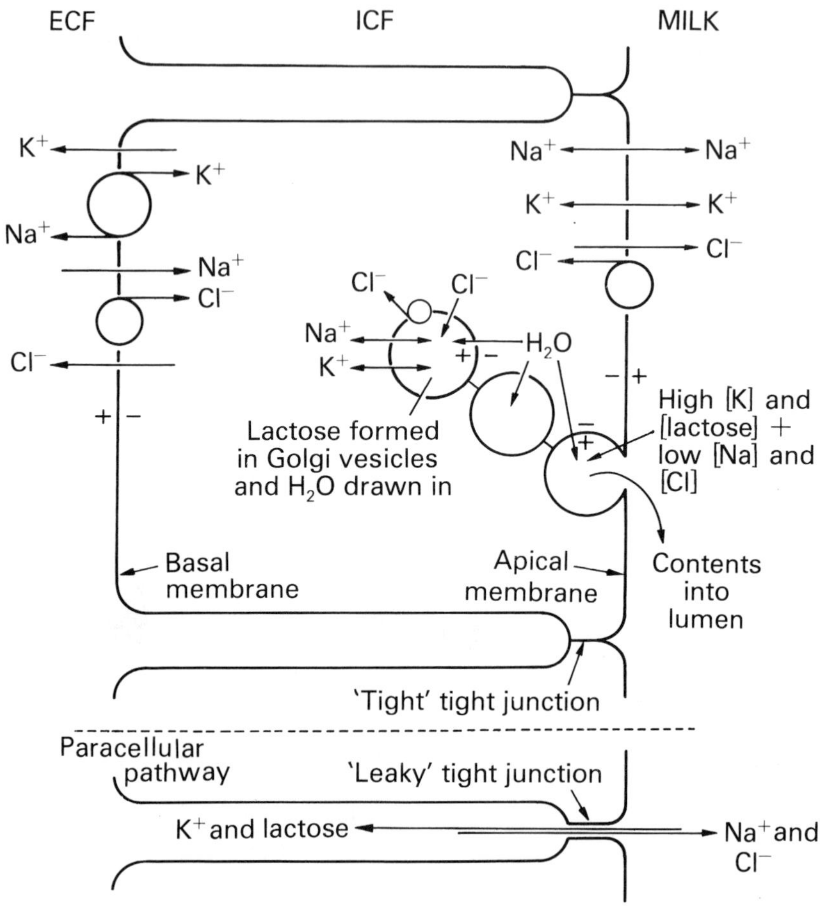

Figure 6.7. Suggested scheme for ion and water movements (and their relation to lactose secretion) between ECF, ICF and milk in the alveolar lumen. Polarities of PDs are indicated by plus and minus signs.
(Source: Peaker, 1978.)

longer time-delay than for protein and monovalent ions (see section 6.4) and indicative of a distinctive secretory process.

It has been suggested that in several tissues TG synthesis takes place within the bilayer leaflets of the ER, so that the droplets formed can either be released on the luminal side of the ER cistern (as in liver cells) or on the cytosolic side, as in the mammary cell. Consequently, the milk fat droplet becomes coated with the outer half of the phospholipid bilayer. On the other hand, some evidence suggests that in mammary cells droplet formation occurs at the cytosolic face of the ER, rather than between the bilayer leaflets. But in either event the droplets enter the cytosol and make their way to the apical membrane, a process quite unlike that in other cells, where fat droplets follow the Golgi pathway and are secreted by exocytosis. Recent EM evidence suggests that during intracellular transport the lipid droplets grow in size by continuous attachment to them of *liposomes*, that is, vesicles containing newly synthesized lipid, surrounded by membrane derived from the ER.

As in the case of secretory vesicle transport, cytoskeletal elements are thought to effect intracellular transfer, but direct evidence for the mechanism involved is lacking.

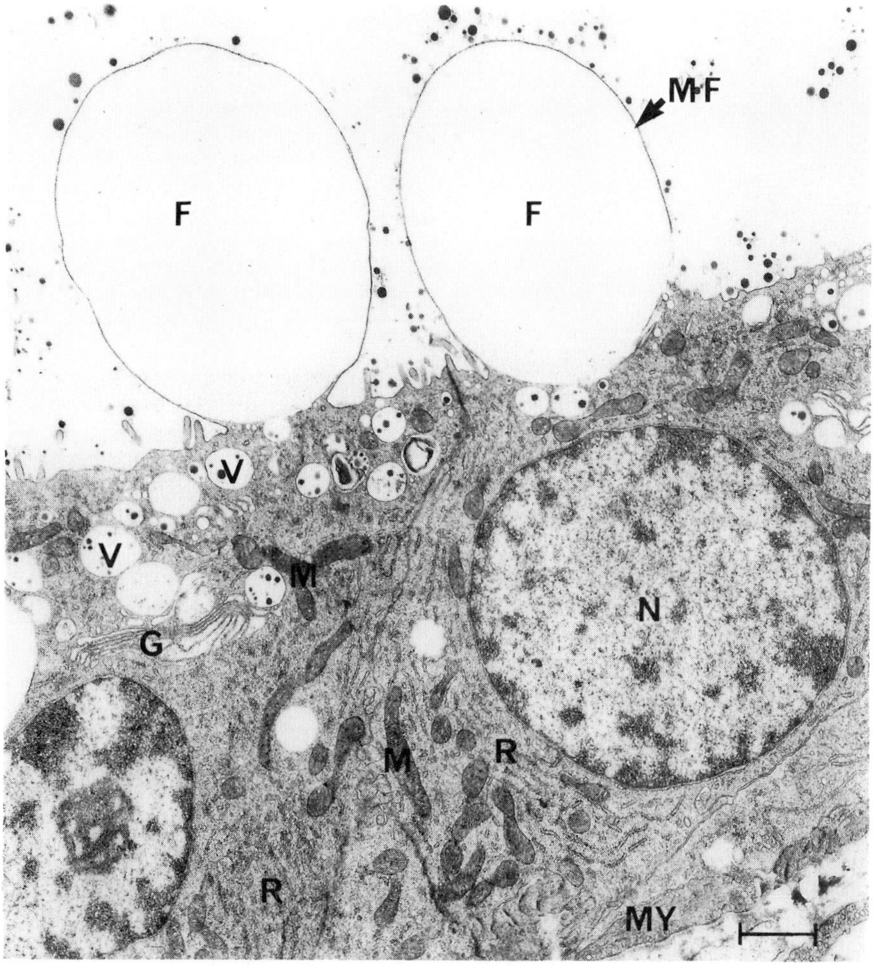

Figure 6.8. Electron micrograph showing secretion of fat droplets by the secretory cells in ovine mammary tissue. (Scale bar, 1.0 μm.) F, fat droplets, MF, milk fat globule membrane; V, secretory vesicles containing protein granules; N, nucleus; G, Golgi apparatus; R, RER; M, mitochondria; MY, myoepithelial cell.
(Courtesy, Dr F. B. P. Wooding.)

There are two main theories which seek to explain the way fat droplets cross the apical membrane (Figure 6.8). That first proposed by Bargmann and Knoop (Figure 6.9a) claims that as the droplet reaches the cytosolic face of the apical

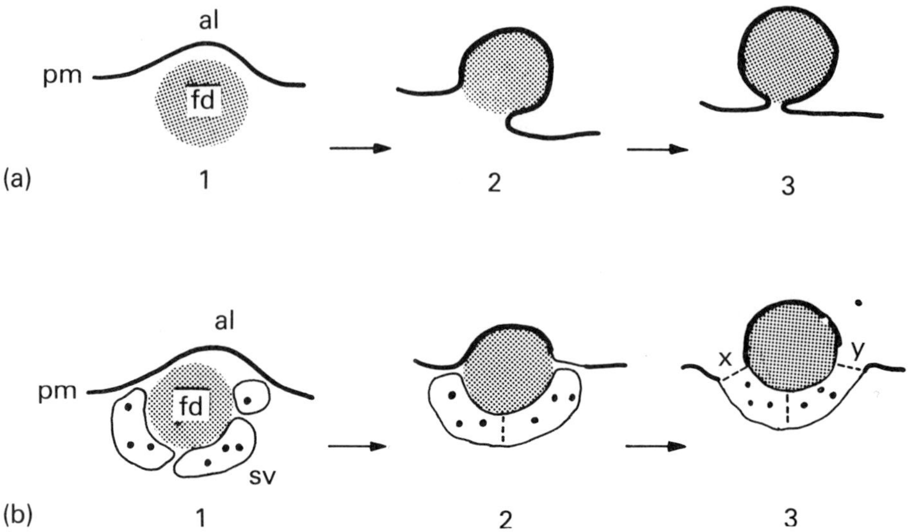

Figure 6.9. Mechanisms proposed for secretion of fat droplets from mammary secretory cells. (*a*) *Bargmann and Knoop* theory: droplet (fd) becomes progressively coated with apical plasma membrane (pm) in stages 1 and 2 until the droplet is eventually 'pinched off' (stage 3). (*b*) *Wooding* theory: secretory vesicles (sv) surround the droplet within the cell and contribute to the MFGM. Release of the droplet occurs when the secretory vesicle fuses with the apical membrane at *x* and *y* (stage 3).
(Source: Mather and Keenan, 1983.)

membrane the latter becomes wrapped around it until it is 'pinched off', surrounded by apical membrane. The second theory, that of Wooding, invokes the participation of secretory vesicles in fat secretion. By encircling the droplet prior to their own fusion with the apical membrane they effect secretion by a process which is, at least partly, exocytotic (Figure 6.9b). The two theories are not necessarily incompatible. What is more problematical is the relative rates at which the two processes proceed, although Wooding denies the occurrence of the pinching-off process as an active mechanism of droplet extrusion.

Electron microscopy suffers from an inherent limitation in studying dynamic processes in that the tissue concerned must be fixed before examination. In theory, a biochemical approach to the problem should provide clearer answers, by allowing comparison of the chemical composition of the MFGM with apical plasmalemma, on the one hand, and with secretory vesicle membrane, on the other. There are, however, significant technical problems in this approach, for example, in the preparation of 'pure' apical membrane fractions. Despite these, there is strong evidence that the MFGM is largely derived from apical membrane rather than secretory vesicle membrane. Thus, the major protein of the MFGM is *butyrophilin*, which immunofluorescence studies show is restricted to the *apical* surface of the secretory cell. Secretory vesicle membranes differ in significant respects from the MFGM.

Whether an intracellular droplet is enveloped by apical or secretory vesicle membrane, an important question concerns the factors effecting fusion of the two surfaces. A theory proposed in the 1960s attributed this fusion to van der Waals (short-range intermolecular) forces generated between the fat droplet surface and the apical plasmalemma, but subsequent EM studies suggested that there is a 10–15 nm electron-dense layer between the droplet and its surrounding membrane, so that the close proximity required for the theory to be applicable does not occur. Some have countered this view by claiming that the major proteins of the electron-dense layer, butyrophilin and *xanthine oxidase*, contain tightly bound fatty acids, which would permit generation of van der Walls forces of sufficient magnitude to effect fat droplet extrusion. Nevertheless, this finding has also been contested and it is clear that, currently, the true nature of the process remains obscure.

In most cases, the spherical fat droplets in the alveolar lumen are surrounded by MFGM (see Figure 6.8). In a few cases, however (approximately 5% of droplets in goat's milk), the droplets have crescents of cytoplasm, sometimes containing ER and Golgi vesicles, attached to them (see Figure 6.1). The numbers of such *signets* are increased by administration of oxytocin to lactating animals, and since this may cause distortion of the secretory cells by vigorous myoepithelial cell contraction (see section 8.2.9) it has been surmised that these mechanical forces lead to premature droplet secretion so that cytoplasmic contents become included in the membrane 'envelope'.

6.8. Membrane flow

In several of the processes described above endomembranes of the secretory cell have been implicated in the transfer of material from one cell organelle to another. Such observations, coupled with chemical analysis of membranes from different organelles, have led to the concept that there is a continuous flow of membrane through the cell, with selective addition and deletion of particular molecules occurring as the membrane proceeds along the route: RER → *cis* Golgi → *trans* Golgi → secretory vesicles → apical plasmalemma → MFGM.

Since apical membrane is *lost* during fat secretion and *gained* as a result of exocytosis of protein and lactose, it might be imagined that coordination of the two processes would ensure a constant apical surface area, but, in fact, the situation seems to be much more complicated. It has been estimated that in cows the addition of membrane from exocytosis may be up to 200 times that lost from the cell in fat secretion. Some of the surplus membrane material may be shed from the cell, but the amounts present in milk are only of about the same order as that in the MFGM (some of which dissociates from the droplet after secretion), so that the bulk of the extra membrane must be *recycled* into the cell. Indeed, membrane-recycling is a widely accepted phenomenon for protein-secreting cells in general.

More precise definition of the processes involved requires kinetic analysis of the time-course of distribution of label in the different membrane compartments following administration of labelled membrane precursors. Few such studies have been performed on mammary tissue, so that the following discussion relates to studies on other types of secretory tissue. Despite problems associated with obtaining labelled membrane fractions uncontaminated with labelled secretory products, it has been possible to trace the passage of radiolabel through the endomembrane system of some tissues and to establish that the half-life of membrane proteins is at least an order of magnitude greater than that of secretory proteins, as would be predicted if membrane recycling were to occur.

A recently advanced theory claims that the Golgi apparatus consists of two distinct, but interrelated organelles. According to this concept, based on the specific distribution of certain enzymes, the *cis* cisternae of the Golgi are related to ER membranes, whereas the *trans* cisternae are related to those of secretory vesicles (see Figure 6.1). While the transfer of secretory proteins from ER to *cis* Golgi involves little selectivity of membrane flow, thereafter only membrane proteins destined to reach the apical surface are transferred to the *trans* Golgi, while others are shuttled back to the ER membrane. Analogously, recycling of the apical proteins is restricted either to the *trans* Golgi cisternae or to degradation in lysosomes. The process is thus equivalent to 'fractional distillation', in that only membrane proteins destined for the apical surface escape being returned to the ER.

Such studies indicate plausible schemes for the involvement of endomembranes in the secretory process, which seem likely to be applicable to mammary cells. But the unique mechanism of milk fat secretion complicates the situation still further. Thus, EM evidence suggests that as fat droplets are secreted there is a clearing of intramembranous particles from the MFGM, so that MFGM and apical membranes are not of equivalent composition. Additional mechanisms would seem to be required for the *selective* removal of such particles, which would otherwise accumulate in the apical membrane.

6.9. Secretion of immunoglobulins

Species-differences in the relative importance of colostral Igs have been discussed earlier (see section 3.3.3).

The IgG of bovine colostrum is exclusively derived from blood plasma. It is presumed that cellular uptake involves binding of IgG molecules, via the Fc fragment, to receptors situated in the basolateral membranes, and it is significant that just prior to parturition there is a sharp increase in the numbers of such receptors showing a high affinity for IgG_1, which is selectively transported into bovine colostrum. The intracellular transport route has not been described with any degree of certainty, but the most likely scheme would seem to involve vesicular transport followed by exocytosis at the apical membrane.

IgA in colostrum is partly derived from intramammary synthesis and partly by accumulation in the gland after being transported in the blood from other sites of synthesis. (The immunological significance of local IgA synthesis is discussed in section 3.3.3.) In either case, IgA molecules are transported into secretory cells across the basolateral membrane by means of a large, membrane-bound form of secretory component (see section 3.3.3), which acts as a recognition site. It is presumed that, following *endocytosis*, the SIgA complex is transported to the apical membrane of the secretory cell where, following cleavage of a portion of the complex, the mature SIgA complex is secreted by exocytosis.

6.10. The rate of milk secretion

Astley Cooper, the eminent British anatomist, claimed, in a book published in 1840, that milk secretion was increased during the act of suckling, in consequence of an increased mammary blood flow. This view persisted well into the twentieth century, with Penlington, in 1915, claiming in a dairying textbook '... milk waiting in the reservoir is small in quantity and poor in quality: the largest and richest portion is manufactured during the actual process of milking.' It is clear that ejection was confused with secretion. But even now we are uncertain whether the secretory process is continuous or intermittent. Even if secretion is continuous the rate may be variable and, indeed, goats milked at exactly 6 h intervals show a slight, but statistically significant, diurnal variation in yield.

More fundamental, however, is the question of whether the process by which milk constituents cross the apical membrane is elicited by a specific stimulus (*stimulus–secretion coupling*, as in the exocrine pancreas) or *deregulated* (as in plasma cells and fibroblasts). There is, for example, strong evidence that Ca^{2+} ions stimulate pulsatile release of zymogen granules from pancreatic acinar cells. If pulsatile release were sufficiently rapid, or uncoordinated throughout the mammary gland, the net result would hardly be distinguishable from continuous secretion in terms of the overall secretion rate. Moreover, there is some autoradiographic evidence that secretion may be discontinuous. Thus, when pulse-labelled with ^{3}H leucine, the cells of individual alveoli appear to be at similar secretory stage, suggesting that there is regulatory communication between contiguous intra-alveolar cells. Gap junctions (see section 2.3.2) may provide the route by which this local synchrony is achieved.

6.11. Conclusion

The reader may be forgiven for feeling that there is much confusion about several of the concepts discussed in this chapter. Certainly, although considerable progress has been achieved as a result of the introduction of techniques such as electron microscopy and PD measurements, the subject exemplifies as well as any the uncertainties which exist at the advancing edge of research. Perhaps because electron micrographs appear to give so clear and unequivocal a picture of cells we are deceived into thinking that the *process* involved should also be visible. In fact, EM photographs are instantaneous 'stills' (prepared by procedures which, in theory, could distort the true picture): in themselves they can only provide insight into dynamic processes by painstaking skill and inspired extrapolation.

Further reading

Berga, S. E. and Neville, M. C. (1985). Sodium and potassium distribution in the lactating mouse mammary gland *in vitro*. *J. Physiol.* **361,** 219–30.

Franke, W. W., Luder, M. R., Kartenbeck, J., Zerban, H. and Keenan, T. W. (1976). *J. Cell. Biol.* **69,** 173–95.

Keenan, T. W., Morré, D. J. and Huang, C. M. (1974). Membranes of the mammary gland, in *Lactation*, Vol. 2, pp. 191–233, eds Larson, B. L. and Smith, V. R. (New York and London: Academic Press.)

Keenan, T. W., Franke, W. W., Mather, I. H. and Morré, D. J. (1978). Endomembrane composition and function in milk formation, in *Lactation*. Vol. 4, pp. 405–36, ed. Larson, B. L. (New York and London: Academic Press.)

Keenan, T. W. and Dylewski, D. P. (1985). Aspects of intracellular transit of serum and lipid phases of milk. *J. Dairy Sci.* **68,** 1025–40.

Keenan, T. W., Dylewski, D. P., Woodford, T. A. (1983). Origin of milk fat globules and the nature of the milk fat globule membrane, in *Developments in Dairy Chemistry*, pp. 83–118, ed. Fox, P. F. (London and New York: Applied Science Publishers.)

Linzell, J. L. and Peaker, M. (1971). Mechanisms of milk secretion. *Phys. Rev.* **51,** 564–97.

Mather, I. H. and Keenan, T. W. (1983). Function of endomembranes and the cell surface in the secretion of organic milk constituents, in *Biochemistry of Lactation*, pp. 231–83, ed. Mepham, T. B. (Amsterdam and New York: Elsevier.)

Neville, M. C., Allen, J. C. and Watters, C. (1983). The mechanisms of milk secretion, in *Lactation; Physiology, Nutrition and Breast Feeding*, pp. 49–102, eds Neville, M. C. and Neifert, M. R. (New York and London: Plenum.)

Patton, S. and Jensen, R. G. (1976). *Biomedical Aspects of Lactation*. (Oxford and New York: Pergamon Press).

Peaker, M. (1978). Ion and water transport in the mammary gland, in *Lactation*, Vol. 4, pp. 437–62, ed. Larson, B. L. (New York and London: Academic Press.)

Peaker, M. (1983) Secretion of ions and water, in *Biochemistry of Lactation*, pp. 285–305, ed. Mepham, T. B. (Amsterdam and New York: Elsevier.)

Saacke, R. G. and Heald, C. W. (1974). Cytological aspects of milk formation and secretion, in *Lactation*, Vol. 2, pp. 147–89, eds Larson, B. L. and Smith, V. R. (New York and London: Acacemic Press).

Wooding, F. B. P. (1977). Comparative mammary fine structure, *Symp. Zool. Soc. Lond.* **41,** 1–41.

7 Endocrine Control of Mammary Growth and Function

7.1 Introduction

Consequent on each successful mating, the mammary gland performs a complete cycle of growth and differentiation. Within that mammogenic cycle three phases are distinguishable: the major growth phase during pregnancy, the secretory phase during lactation (which in some species, at least, also includes continued growth) and the phase of involution (Figure 7.1). Since mammary regression is sometimes incomplete between successive lactations, the starting point of each cycle may be in advance of that at the previous conception, so that in such cases one might appropriately speak of a mammogenic 'spiral'.

Ever since the beginning of the century, when the importance of hormones became apparent, the hormonal control of mammary growth and function has been a subject of intense scientific curiosity. But in science, in which progress depends largely on general conceptual advances, early attention to problems is no guarantee of their early solution, and the history of attempts to define the hormonal control of mammogenesis, lactogenesis and galactopoiesis is a maelstrom of theories, countertheories and changing emphases. This subject in particular is bedevilled with interspecies variations; it seems that the late arrival of the mammary gland on the evolutionary scene has resulted in a multiplicity of hormonal agencies through which control might be achieved.

In 1900, it was reported that removal of an animal's ovaries led to mammary regression, which could be prevented if the animal received an ovarian transplant. The availability of synthetic oestrogens in the 1930s led to the demonstration that they could induce mammary ductal growth when administered to several species, whereas in combination with progesterone they stimulated full lobulo-alveolar development. It was thus inferred that these ovarian hormones were

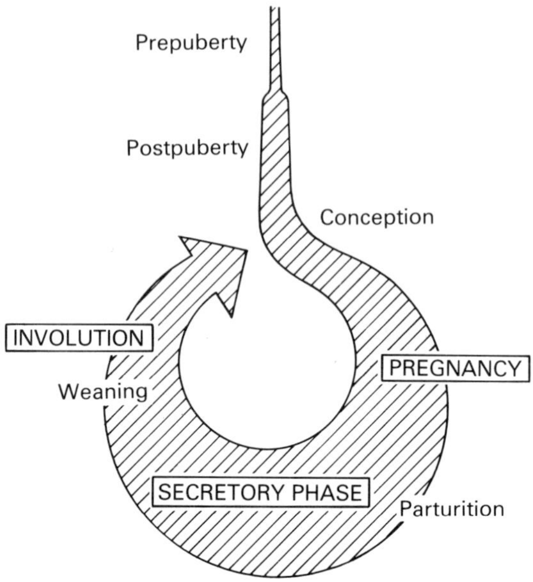

Figure 7.1. Mammary growth during the lactation cycle: a generalized scheme. Changing width of the arrow represents the changing size of the gland during the cycle. (Source: Knight, 1984.)

responsible for the mammogenesis of pregnancy by acting directly on mammary tissue. The mammogenic role of placental hormones, first proposed in 1905, received little subsequent investigation for many years, and only became clearly established in the 1960s.

In 1928, Stricker and Greuter showed that pituitary gland extracts elicited milk secretion when injected into pregnant rabbits, and when subsequently a purer extract was found to have the same effect, the active principle was designated *prolactin*, since its identity was distinct from those of other known pituitary hormones.

The interplay between the roles of pituitary gland, placenta and ovaries was also found to be modulated by hormones secreted by other endocrine glands, so that description of the overall control of mammary function came to be seen as a highly complex problem. Nevertheless, in recent years, largely as a result of the introduction of rapid and increasingly sensitive methods of investigation and the accumulation of data from many different species, general principles have emerged which constitute considerable progress in our understanding of the mechanisms involved.

In this chapter the problem will be discussed at three levels: firstly, the morphological development of the gland; next, endocrine control of cell growth and differentiation; and finally, the endocrine control of milk secretion. Chapter 8 deals with the neural control of galactopoietic hormone release and Chapter 9 pursues the question further by examining the cellular and intracellular actions of these hormones and other regulatory factors.

7.2. Phases of mammary development

7.2.1. Methodology

Before discussing the pattern and extent of mammary development we need briefly to consider methods used to assess it. Several methods are available but they are not all applicable to all species, or at all times in the cycle.

(a) In animals with large pendulous glands, such as goats, volume may be measured by water displacement, using the Archimedes principle.
(b) In cases where it is ethically and economically acceptable, glands may be surgically removed, weighed and assessed for histological and chemical structure.
(c) An index of cell numbers may be derived by estimation of total DNA (DNA_t) of removed glands, on the assumption that the DNA_t of somatic cells is constant: (such estimates do not distinguish between parenchymal and stromal cells unless chemical analyses are also performed).
(d) Biopsies (small pieces of tissue removed under anaesthesia) may be taken from glands of large animals to allow calculation of DNA_t when combined with gland volume measurement (method (a)).

The main problem with assessing mammary development is that all the really useful methods involve invasive procedures which cannot be applied to humans and are expensive to use in large domesticated species. Consequently, the more-detailed information available relates to laboratory rodents which can be used in sufficient quantity to obtain statistically significant data.

7.2.2. Prenatal growth

The glands make their first appearance in the embryo. Two raised lines of ectoderm (*milk lines* or *crests*) appear on the ventral surface, one each side of the midline, running from the axilla to the inguinal region. Subsequently, at positions which vary with species, and correspond to the gland positions in the mature animal, nodules of cells appear on the line and then sink into the dermis to become *mammary buds*. Further development involves the buds elongating into cords, which become hollow ducts lined with cuboidal cells, together with cellular proliferation at the surface to form rudimentary nipples.

Early development is not due to cell proliferation but to morphogenetic movement of cells. Subsequently, the formation of primary and secondary ductal sprouts *does* involve cell division. The growth rate at this stage is similar to that of the body as a whole (*isometry*: a growth rate faster than that of the whole body is termed *positive allometry*). *Sexual dimorphism* is often apparent: for example, in male rats and mice, duct growth is inhibited and nipples are usually absent.

Stromal tissues also develop in the fetus, arising from the mesenchymal layer surrounding the rudimentary gland. The mesenchyme is the source of the mammary fat pad, which is of considerable importance in providing structural support and regulatory control of the developing parenchyma.

7.2.3. *Development in the juvenile*

At birth the mammary gland generally consists of a few primary and secondary ductal sprouts, terminating in a diminutive nipple and encapsulated in a fat pad. Prior to puberty, growth is essentially isometric and is confined to modest extension of the duct system into the fat, which itself extends somewhat by cellular hypertrophy.

In most species a phase of allometric growth begins just prior to onset of the ovarian cycles. Ductal, rather than lobuloalveolar growth is stimulated, but with succeeding cycles the growth apparent in the follicular phase is followed by regression during the luteal phase, so that the *net* gain is slight. The mammary development at puberty with which about 50% of us are most familiar (and the rest, hardly less so) is, in fact, highly atypical: the pubertal growth of the human breast is largely due to fat deposition, which does not occur in other species, but is an important element in human libido (see section 1.8.2).

In certain species, post-pubertal mammogenesis takes the form of extensive lobuloalveolar growth in addition to ductal growth, and results in the condition of *pseudopregnancy*. This may occur spontaneously, after ovulation, as in dogs, or be induced by an infertile mating, as in rabbits and ferrets.

7.2.4. *Mammogenesis in pregnancy*

In most species, however, major lobuloalveolar growth only occurs during pregnancy. DNA_t, and hence gland volume, increases markedly and the histological appearance of the gland is transformed (Figure 7.2) as the lobules grow into the stroma, displacing adipose tissue. The final stage of lobuloalveolar development is called *terminal differentiation*.

In rats and mice, mammary DNA_t increases exponentially during pregnancy, the cell population doubling-time being about 6 days in mice and rather longer in rats; as a generalization, the doubling-time is roughly inversely related to gestation length.

7.2.5. *Growth during lactation*

In rats and mice the exponential rate of parenchymal growth evident in pregnancy continues for about 5 days post partum, representing a significant proportion of the total mammary growth (about 40% during the first lactation of rats). It has been generally believed that in dairy ruminants growth is essentially complete at the time of parturition, but this assumption has recently been called into question by results for goats showing an approximately 20% increase between the first and third week of lactation. In either event, the increases in milk yield in early lactation exceed cell proliferation rates, so that, since yield is the product of cell number and secretory activity per cell, it would be anticipated that the latter increases. The increase in mammary RNA:DNA ratio (see section 4.10) in early lactation is consistent with this assumption.

In most species, following the attainment of peak milk yield early in lactation,

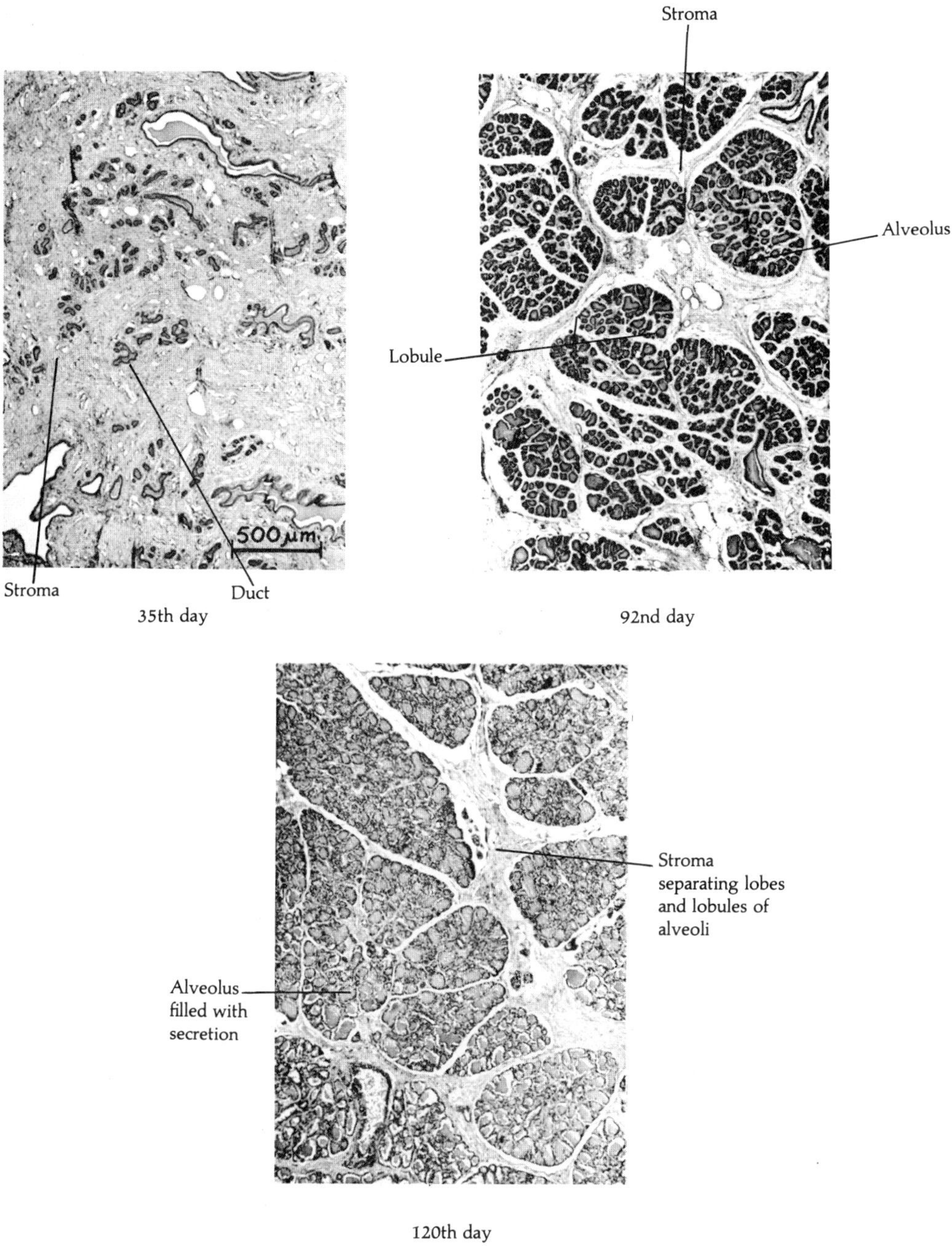

Figure 7.2. Histological sections of goat mammary tissue at three stages of pregnancy (total length approximately 150 days): 35th day, showing small ducts scattered in stroma; 92nd day, showing developing lobuloalveolar tissue; 120th day, showing almost full lobuloalveolar development and a much reduced stroma.

(Source: Cowie, 1971.)

the subsequent decline in yield is more gradual, probably due to both the decreased DNA_t and decreased RNA:DNA ratio.

7.2.6. *Mammary involution*

Three types of involution have been defined:

gradual involution, the progressive regression observed in the lactating gland following peak yield (see section 7.2.5)
initiated involution, due to sudden cessation of milking (for instance, as would occur on death of the young), and
senile involution, regression due to ageing.

The type which has been studied most extensively is initiated involution, where abrupt termination of milking or suckling produces exaggerated effects. In goats, milk secretion is almost completely suppressed after 3 days, while in rats the period is only about 10 h. The initial distension of the gland with accumulated milk subsides within a few days and the gland perceptibly shrinks.

Involution is essentially an autolytic process, with lysosomes playing the major role, while an influx of phagocytic cells aided by disruption of the tight junctions (see section 2.3.2) completes the degenerative process. Progressively, parenchymatous tissue is replaced by connective and adipose tissue.

Equivalent changes, occurring much more slowly, accompany the normal weaning process. The increasing prominence of the paracellular route leads to the characteristic changes in milk composition, previously described (see section 6.6).

7.2.7. *Gland structure between successive lactations*

The interval between successive lactations (the *dry period*) varies markedly between, and within, species. For example, in dairy cows, which are usually pregnant in the later stages of lactation, the dry period is about 2 months, whereas in women it may amount to several years. It is a matter of some interest to know whether cells 'carry-over' from one lactation to the next, what the secretory consequences of such persistence are, and the extent to which carry-over is determined by the interval between lactations. But few studies have addressed this particular problem. In cows given a 'unilateral' drying-off period, the subsequent post partum milk yield of the continuously milked udder-half was only 70% of that which had been dried-off, suggesting the importance of intramammary factors. Experiments with rats, employing ^3H thymidine administration (which is largely incorporated into DNA), suggested that there was appreciable persistence of mammary cells from one lactation to the next, a finding which is consistent with the observation that milk yields of dairy animals increase with parity for several successive lactations.

7.3. Control of mammogenesis

7.3.1. Investigative procedures

Most of our knowledge of the control of mammary growth relates to rodents and ruminants. Interspecies variations in the hormones involved are marked, making it difficult to generalize with confidence. Of necessity, phrases like 'in many species' will recur frequently in the following discussion.

Finer definition of the factors involved has been possible as techniques of investigation have evolved. Early experiments employed extirpation of endocrine glands (*endocrinectomy*), followed by injection of extracts of the glands (*replacement therapy*) to assess the extent to which the induced physiological lesions could be corrected. Subsequently, the active principles of the extracts (hormones) were chemically isolated and their effects on endocrinectomized and normal animals investigated. The introduction of bioassay procedures for measuring the concentrations of hormones in blood was, more recently, augmented by rapid and more sensitive radioassay techniques, and latterly, with the realization that hormones act by binding to *receptors* in the cells of their target organs (see section 9.2), *radioreceptor assay* methods have been developed.

7.3.2. Fetal development

Early embryonic development of mammary tissue does not appear to be influenced by hormones, but is highly dependent on mesenchymal factors. Thus, mammary tissue from 12 day rabbit embryos, cultured *in vitro*, only developed into a primary mammary bud in the presence of mammary mesenchyme; indeed, the latter induced bud formation is non-mammary epithelium. The remarkable potency of the mesenchymal stimulus is demonstrated by the fact that it elicited development of bud-like structures when cultured with epidermal tissue from a 6 day *chick* embryo, although by day 8 such tissue is committed to feather formation and did not then give a response.

Later stages of fetal development are subject to fetal endocrine influences. *Androgens* (probably, chiefly *testosterone*) are instrumental in suppressing mammary development in male rodents, but the sensitive period is very short, for example, from days 13 to 15 of fetal life in mice. The loss of sensitivity is probably important because the female fetal adrenal gland later secretes testosterone.

The capacity to respond to hormones is perhaps most readily evident in the observation that at birth young of several species show a precocious mammary development, associated with secretory activity; for example, human babies (where the secretion has been called 'witch's milk'), rabbits and bovine calves. This activity appears to be due to transient increases in the fetal blood concentrations of lactogenic hormones.

In vitro culture of fetal mammary tissue from certain species shows a mammogenic response in the presence of insulin, adrenal steroids and prolactin, the same combination which proves effective for adult mammary tissue (see section 9.4.2). The influence of ovarian steroids is less certain, but the potentiating effects of oestrogen and progesterone on tumour growth provide a signal warning of the possible hazards of their therapeutic use in pregnant women.

7.3.3 *Puberty*

The extension of the duct system in the follicular phase of the oestrous cycle depends on oestrogen secretion; there is virtually no lobuloalveolar growth in most species. The cyclic release of oestrogen produces an interrupted pattern of ductal growth (see section 7.2.3). Some evidence suggests the importance of *epidermal growth factors* or a specific mammary *growth factor* (probably both peptides) in mammogenesis, and it is possible that such factors are secreted by other organs, for example, kidney or uterus.

Any lobuloalveolar development in the virgin animal depends on the combined effects of progesterone and oestrogen, but these hormones are only effective in the presence of pituitary hormones (see section 7.3.4). In animals with a prolonged luteal phase of the oestrous cycle, the secretion of progesterone promotes extensive lobuloalveolar development.

7.3.4. *Pregnancy*

Let us first consider the hormonal 'environment' of pregnancy. In all species, maintenance of pregnancy is initially dependent on the secretion of progesterone from the *corpus luteum*. Its blood concentration is elevated in most species, but supply to target organs is in some cases ensured by reduced metabolic clearance rates, due to increased amounts of a specific binding protein in blood. In most species also, oestrogen concentrations increase during pregnancy. For both of these ovarian steroids, the fetoplacental unit becomes a major site of synthesis in the later stages of pregnancy in many species, so that ovariectomy affects neither pregnancy nor mammogenesis. The limited data available on blood concentrations of adrenal steroids during pregnancy show no consistent inter-species trends.

Of the anterior pituitary hormones, prolactin concentrations in blood generally increase during pregnancy, while those of growth hormone (GH) show little change. In dairy ruminants, prolactin concentrations show seasonal variations, being positively correlated with daylight length and intensity, and with environmental temperature. By contrast, in some species there is a marked increase in blood concentrations of *placental lactogen* (a molecule the structure of which is related to that of prolactin) during pregnancy. For example, in goats its concentration begins to rise at mid-pregnancy and may reach 1 $\mu g\ ml^{-1}$ plasma prior to parturition (Figure 7.3a). Placental lactogen also seems important in rats and mice, but its role is uncertain in cows.

What effects do such hormonal changes have on mammogenesis? According to current theories, the sex steroids (from ovary or placenta) are important in pregnancy both by stimulating prolactin release and by direct action on the mammary parenchyma in synergy with pituitary hormones. Oestrogen, in association with pituitary and adrenal hormones, promotes ductal development, while progesterone is additionally required for lobuloalveolar development. Later in

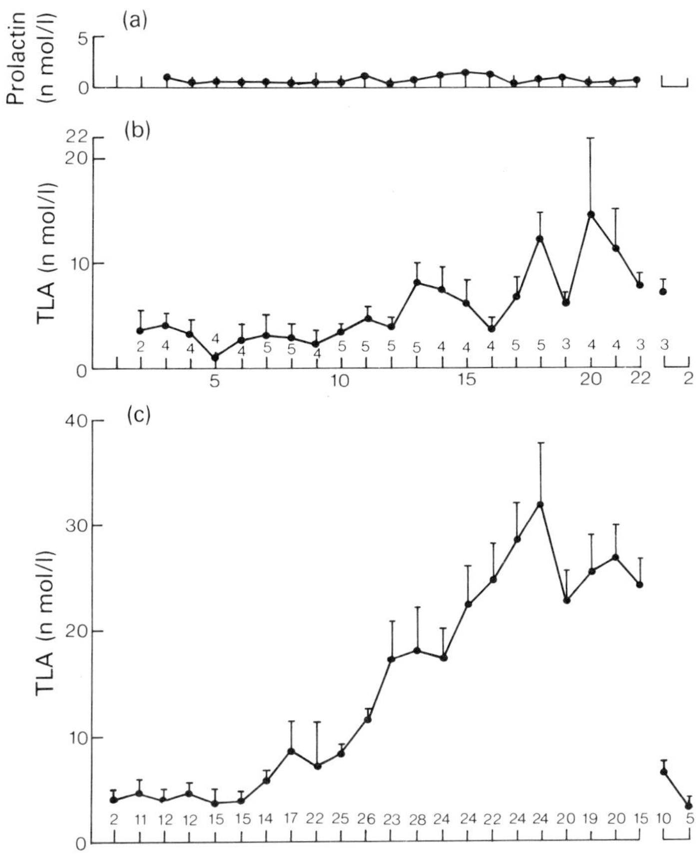

Figure 7.3. Lactogenic hormones in the blood of goats during pregnancy. (*a*) Prolactin in blood plasma of six goats (measured by specific radioimmunoassay); standard errors of means are too small to be shown. (*b*) Total lactogenic activity (TLA), i.e. mostly placental lactogen, in nmolar equivalents of ovine prolactin, in goats carring one fetus (measured by radioreceptor assay); (*c*) TLA in goats carrying two fetuses. Vertical bars represent standard errors of mean.
(Source: Hayden *et al*, 1980.)

pregnancy, the fetoplacental unit may take over the role of the pituitary, placental lactogen replacing or augmenting the action of prolactin.

The mammogenic role of placental lactogen in goats is nicely illustrated by the fact that there is a significant positive correlation between its blood plasma concentration in late pregnancy (itself correlated with the number of fetuses *in utero*—Figure 7.3b and c) and the subsequent mammary gland weight and milk yield.

Thus, despite their denominations, 'prolactin' and 'placental lactogen', these protein hormones have pronounced *mammogenic* effects: indeed, an alternative name for placental lactogen is *chorionic mammotrophin*. Moreover, it is likely, contrary to earlier views, that these hormones play the dominant role in mammogenesis, since both *in vivo* and *in vitro* experiments indicate that mammary development *requires* the presence of either or both of them, but may occur in the absence of ovarian steroids.

7.3.5. *Growth during lactation*

The mammary growth in early lactation, which is firmly established for rodents, and probably also occurs in ruminants (see section 7.2.5), is clearly associated with a quite different hormonal *milieu*. In almost all species, parturition is associated with an abrupt fall in progesterone concentrations and, in those species in which it is present, in placental lactogen concentrations. How then is growth maintained when essential elements of the mammogenic complex of pregnancy are withdrawn?

We cannot be certain of the answer, which in any case is likely to be species-variable, but there are some suggestive clues. In all species studied, blood prolactin concentrations increase at parturition, as, frequently, do those of GH: and suckling stimulates the release of both of these hormones (see section 8.3). Furthermore, *local factors* seem to be important. Thus, the mammary gland appears to be able to synthesize ovarian steroids (such as *oestradiol-17β* in the late-pregnant goat mammary gland) and *prostaglandins*. But the situation becomes potentially even more complex when we consider the possible involvement of locally produced growth regulators. These might be either the growth factors previously mentioned (in section 7.3.3) or mammary *chalones*, 2000–3000 molecular weight substances extracted from mammary tissue, which specifically inhibit mammary development when injected into other animals. The way in which these different factors interact is highly problematical, and presents a formidable challenge to those engaged in research.

There is a dearth of information on the factors affecting the decrease in mammary DNA_t and RNA:DNA ratio following peak milk yield. We do not know to what extent cell loss is a random process, consequent on the mechanical 'wear and tear' of repeated milk ejection phases or a low-level autolytic process, analogous to that seen in initiated involution (see section 7.2.6). In either case, it might be facilitated by decreasing concentrations of galactopoietic hormones or reduced sensitivity to their effects. Progressively less efficient emptying of the alveolar lumina may exacerbate the process via accumulation of local factors (see section 4.9.1).

Figure 7.4 summarizes the changes in mammary growth and composition throughout the life-cycle of rats and mice. It must be stressed that the scheme is *tentative* only and may well prove to be inaccurate in detail. One aspect of questionable validity is the division of parenchyma into lobuloalveolar and ductal components, since there is known to be appreciable overlap.

7.3.6. *The mammary gland as an endocrine organ*

Several recent studies with goats have suggested that, in addition to being a target organ for many hormones, the mammary gland is also a site of hormone

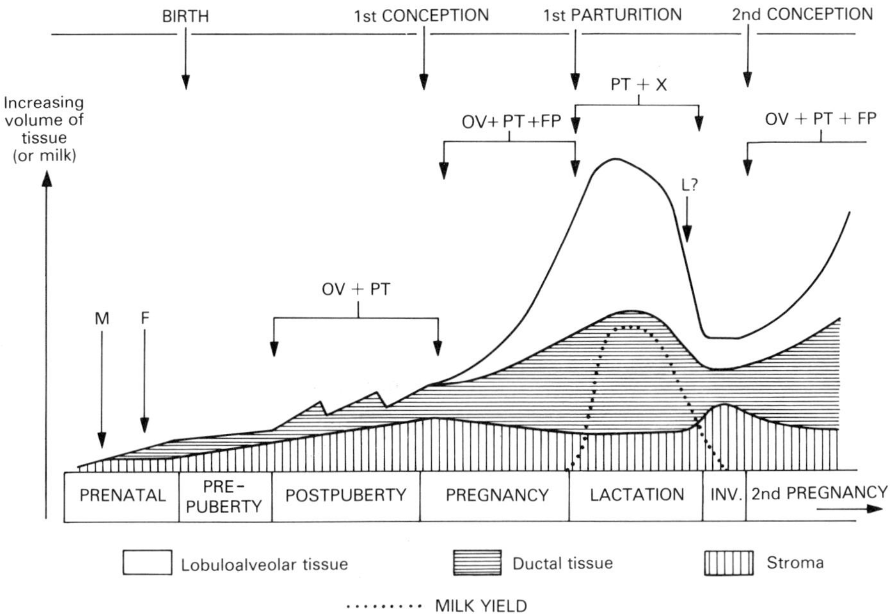

Figure 7.4. Schematic representation of mammary growth in rats and mice. INV, involution; M, mesenchymal factors; F, fetal endocrine factors; OV, ovarian hormones; PT, pituitary hormones; FP, fetoplacental hormones; X, other humoral factors (identity uncertain); L, local factors (identity unknown).

production. The demonstration that the glands in pregnant animals produce progesterone from pregnenolone was followed by reports that it also synthesizes oestradiol 17β and *prostaglandin* $F_{2\alpha}$ ($PGF_{2\alpha}$). Whereas mammary production of progesterone and $PGF_{2\alpha}$ may not significantly increase their blood concentrations (in the latter case because of rapid inactivation in the lungs), that of oestradiol-17β is thought to be largely responsible for the rise in its blood concentration in late pregnancy. We must, therefore, consider the mammary gland as possessing both exocrine *and* endocrine functions.

Indeed, the gland seems to be important in regulating reproductive activity in ways which are, at present, only poorly perceived. For example, *mastectomy* (gland removal) in goats (a) reduces the length of the oestrous cycle, (b) extends the duration of behavioural oestrus, and, possibly, (c) reduces the length of gestation, in animals studied 1 year post-operatively. Such observations are consistent with the postulated endocrine function of mammary tissue, but while some of the humoral factors involved may be 'recognized' hormones, it is possible that others secreted are specific to the mammary gland.

7.4. Hormonal control of lactation

7.4.1. Lactogenesis

In the above account of mammogenesis, only oblique references were made to milk secretion. In this section our concern is with the physiological and endocrine changes associated with this process.

By any standards, the 'switching-on' of milk secretion is a precipitous event, a copious flow of milk being achieved within hours, or at most, days, of parturition. But it is not instantaneous; so major a shift in the body's metabolic activity could hardly be so. It is useful to distinguish two stages of lactogenesis: stage I, at the beginning of which specific milk components first make their appearance in mammary tissue, and which persists up to the time of parturition; and stage II, the usually much briefer period at, or shortly after, parturition, when the copious flow of milk begins. Very few species have been extensively studied at lactogenesis, but even so it is evident that wide interspecies differences exist. For example, in goats, with a 5 month pregnancy, stage I begins about 3 months prepartum (Figure 7.5), whereas in rats, with a 3 week pregnancy, stage I begins only about 30 hours prepartum. In several species, including human females, cows and sheep, it is possible to detect the specific milk protein, α-lactalbumin, in blood plasma from about mid-pregnancy, presumably following its diffusion across the leaky mammary epithelium.

In goats there is a progressive increase in the lactose and protein concentrations and decrease in the Na^+ and Cl^- concentrations in fluid (termed *precolostrum*) removed from the glands throughout stage I. At parturition, coincident with the onset of stage II, not only is milk flow rapidly enhanced but the glands absorb increased quantities of metabolic substrates from the blood as a consequence both of increased AV differences (see section 4.5) and increased rate of mammary blood flow (Figure 7.6). A marked transition in secretory composition, from colostrum to milk (in which lactose and potassium concentrations are much higher and those of immunoglobulins and serum proteins much lower) generally occurs over a period of a few days (see Figure 7.5). Stage I may thus be characterized as due to gradual chemical and morphological changes (such as closure of tight junctions), and stage II as the result of abrupt cardiovascular, metabolic and secretory changes.

7.4.2. Endocrine correlates of lactogenesis

Stage I of lactogenesis is easier to study in large animals, in which it is possible to remove small quantities of secretion during pregnancy without unduly disturbing the course of events. In laboratory rodents this is not practicable and determination of the chemical content of the mammary tissue necessitates killing the animals; this precludes accurate timing of events because the exact time at which parturition *would* have occurred is uncertain. In sheep and goats the appearance of lactose or α-lactalbumin in mammary secretions roughly coincides with the appearance of elevated concentrations of placental lactogen in maternal blood, and it has been speculated that the two events are causally related.

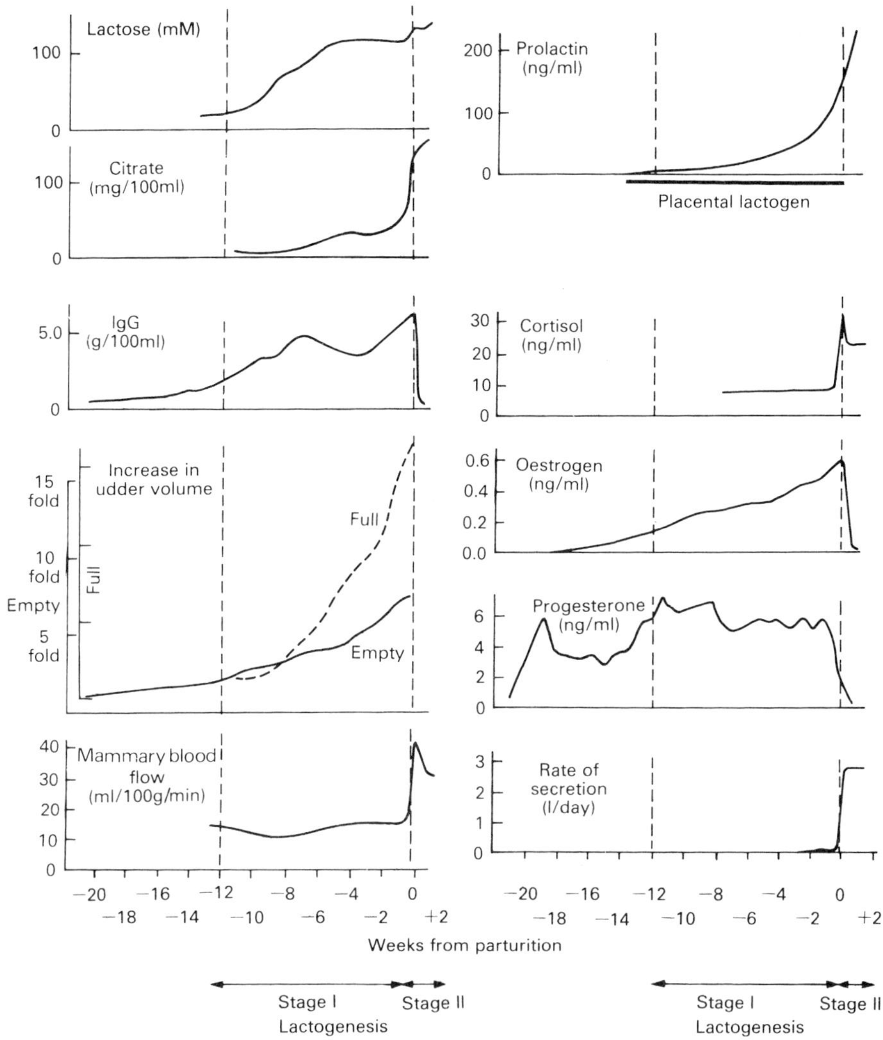

Figure 7.5. Lactogenesis stages I and II in goats. Changes in hormone concentrations in blood, udder volume, MBF, milk secretion rate and concentrations of IgG, lactose and citrate in mammary secretions.
(Source: Fleet *et al.*, 1975.)

At parturition the mammary gland is exposed to the abrupt change in the hormonal *milieu* already discussed (see section 7.3.5). Since the nutritional needs of the neonate become even greater, at a time when its umbilical link is severed, it would be anticipated that the endocrine control of parturition and lactogenesis would be closely integrated. In broad terms, two types of hormonal mechanism can be envisaged: that is, lactogenesis stage II results from (a) *release* from inhibition by a factor(s) which declines in activity at parturition, and/or (b) *stimulation* by a factor(s), the activity of which is enhanced at parturition.

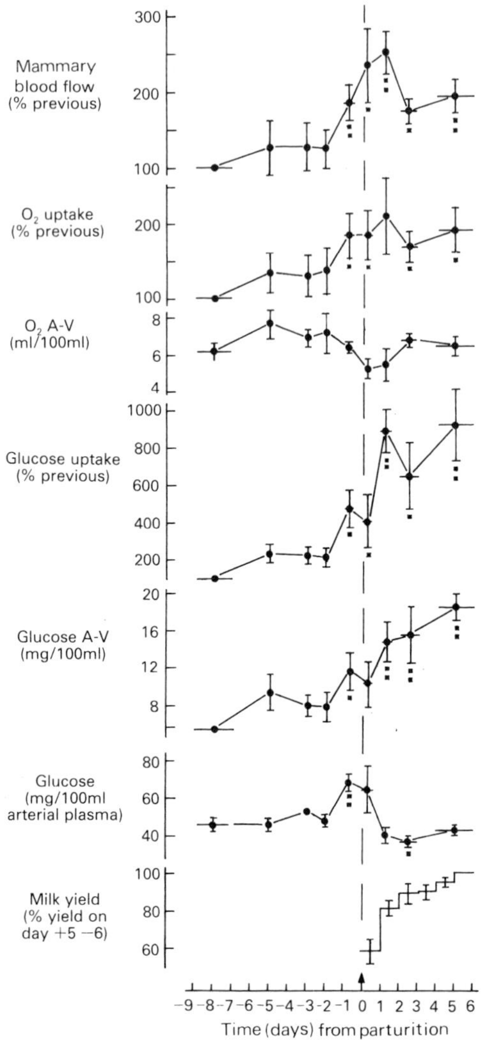

Figure 7.6. MBF, oxygen AV difference and mammary uptake, glucose AV difference and mammary uptake, arterial glucose concentration and milk yield during pregnancy and early lactation in six goats. Horizontal bars represent periods over which data were bulked. ■$p<0.05$; ■■$p<0.01$, with respect to values 7–9 days *prepartum*.
(Source, Davis *et al.*, 1979.)

Currently, the most widely accepted theory is that it is the dramatic fall in progesterone concentration, usually beginning just prior to parturition, which releases the mammary epithelium from an inhibitory state. This theory is consistent, for example, with the observation that the lactogenic response which follows ovariectomy of pregnant rats is prevented by giving them injections of progesterone. Progesterone withdrawal has thus been designated the *lactogenic trigger*.

In most species, blood concentrations of prolactin and adrenal steroids increase at parturition, and, since early experiments with endocrinectomized rats, and later organ culture experiments, suggested their importance in initiating milk secretion, it has been proposed that they may act as a positive stimulus to lactogenesis. The evidence does not, however, sustain that view. Prolactin is undoubtedly important in a 'permissive' sense, since inhibiting its release with *bromocriptine* (see section 8.3.2) prevents lactogenesis, but elevation of its blood concentration does not seem to elicit milk secretion.

Thus, as a generalization, progesterone, which is so important in stimulating lobuloalveolar development in pregnancy, seems also to be the principal factor holding both parturition and lactogenesis in abeyance. Its inhibitory activity is not absolute, since prolactational factors permit a minimal degree of secretory activity in stage I, but only when progesterone concentrations begin to decline are its defences breached, allowing the cellular machinery which has stood 'at ready' for so long to move into full activity.

7.4.3. The stimulus for progesterone withdrawal

If progesterone withdrawal is the lactogenic trigger, what pulls the trigger? There is no single, universally applicable reply to this question, but in goats and sheep, the answer, in the best traditions of 'whodunnit' novels, is a seeming innocent—the fetus. The stimulus for both lactogenesis and parturition originates in the fetal hypothalamus, the activity of which leads to elevation of fetal cortisol concentrations. Cortisol enhances the activity of certain placental enzymes, resulting in increased production of oestrogens and, in the sheep, to a fall in blood progesterone concentrations, since the placenta is the main source of this hormone. In goats the elevated oestrogen concentrations stimulate the uterine production of $PGF_{2\alpha}$ which reaches the ovary via the circulatory system, and, in consequence of its luteolytic property, leads to a decline in progesterone concentrations in blood.

In other species, different mechanisms may operate. For example, in rabbits, the fetus is not involved, lactogenesis resulting from changes arising in the maternal hypothalamus.

7.4.4. Local factors in lactogenesis

Lactogenesis stage II is dependent on factors arising in the mammary gland as well as on systemic factors. This aspect of control has been studied most extensively in goats.

In addition to its uterine production, $PGF_{2\alpha}$ is also synthesized in the mammary gland. In late pregnancy it is secreted into the mammary venous blood, but beginning between 3 and 5 days prepartum two dramatic changes occur: (a) the destination of $PGF_{2\alpha}$ secretion is switched from venous blood to alveolar lumina (that is, across the apical rather than the basal surface of the epithelium), and (b) $PGF_{2\alpha}$ is largely metabolized to *13,14-dihydro-15-keto* $PGF_{2\alpha}$ $DHKPGF_{2\alpha}$). It seems likely that $PGF_{2\alpha}$ is an inhibitor of milk secretion, so that at parturition it must be removed or inactivated. Production of $DHKPGF_{2\alpha}$ provides the latter means of releasing the mammary cells from inhibition, while the removal of milk by the suckling young, immediately following parturition, continues the process.

It is possible that the changes in mammary secretion and metabolism of $PGF_{2\alpha}$ are causally related to the fall in placental lactogen concentration which occurs in goats at this time, but the concurrently increased mammary production of oestradiol 17β (see section 7.3.6) may also be involved.

Lactogenesis in goats, and probably also in other species, thus involves a complex interaction between fetus and mother (Figure 7.7), which operates both

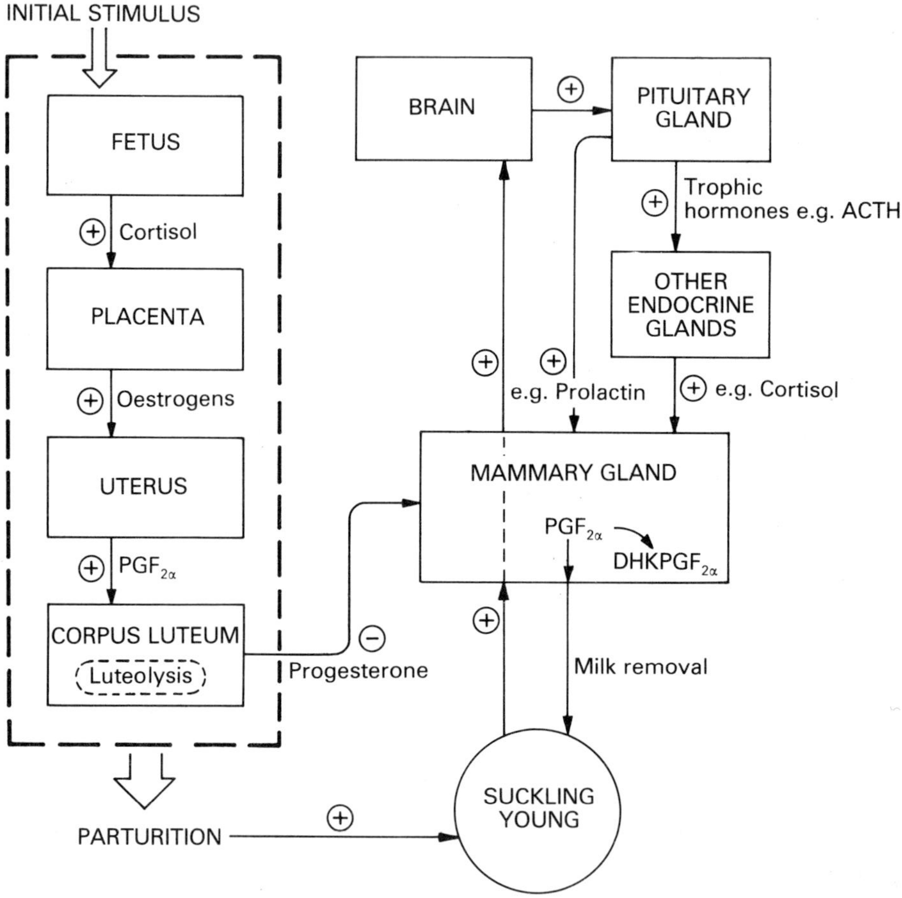

Figure 7.7. Endocrine control of lactogenesis in goats: a summary. The scheme is based on results from several laboratories. +, increased concentrations or activity: −, decreased concentration. Events occurring *prepartum* are enclosed in the large dashed box.

before and after parturition. In addition to eliciting secretion of lactogenic hormones, like prolactin and adrenal steroids, suckling also exerts local control over secretion by removing $PGF_{2\alpha}$ from the gland. If the number of glands exceeds the number of suckling young, the unsuckled glands will clearly regress rapidly as unremoved $PGF_{2\alpha}$ continues to exert its inhibitory effect.

7.5. Established lactation

It is customary to refer to the succeeding phases of established lactation as *early* (that is, when yield is still increasing), *peak* (the phase of maximal yield, not always as transient as the term might suggest) and *late* lactation (an imprecise phase prior to involution). We have noted the rough parallel between changes in mammary DNA_t and RNA:DNA ratio on the one hand and milk yield on the other (see section 7.2.5). Are there clear hormonal correlations with such changes in secretory activity?

Early experiments involving endocrinectomy and replacement therapy established the identity of galactopoietic 'complexes' of hormones for several species, which exhibited the, by now anticipated, interspecies variation in their composition. *Hypophysectomy* (pituitary gland removal) causes a rapid inhibition of milk secretion in all species, but whereas in rabbits yield can be completely restored by injection of prolactin alone, rats require prolactin and adrenal steroids, while goats (Figure 7.8) and sheep also need GH and thyroid hormone. It is, of course, possible that some of these hormones have no specific galactopoietic action but are required to rectify a general metabolic deficiency resulting from hypophysectomy.

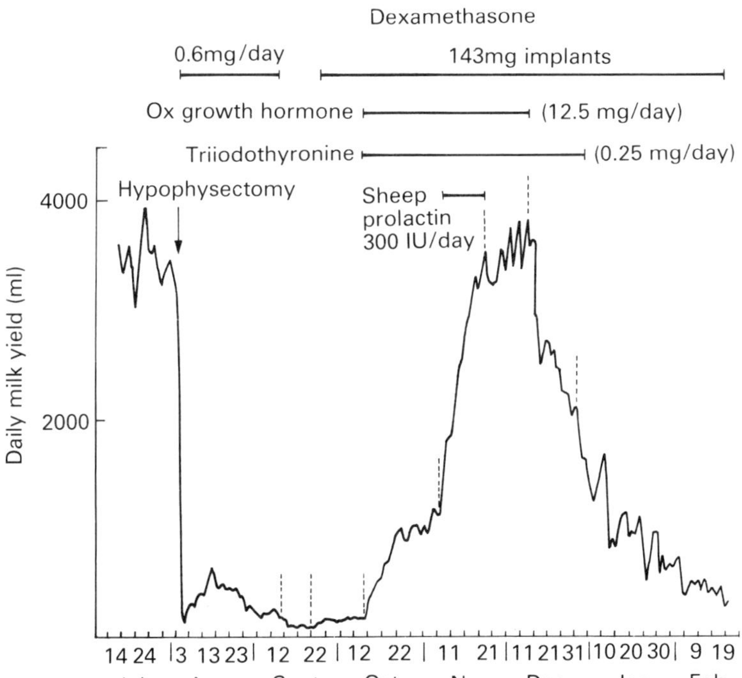

Figure 7.8. Milk yield of a goat which was hypophysectomized during lactation and subsequently treated with hormones.
(Source: Cowie, 1969.)

More recently, use of the prolactin release inhibitor, bromocriptine (see section 8.3.2), has confirmed the importance of this hormone at lactogenesis in all species studied, but revealed that in established lactation suppression of prolactin release does not significantly affect milk yield in cows and goats, although it does continue to have a marked inhibitory effect in other species, such as rabbits (Figure

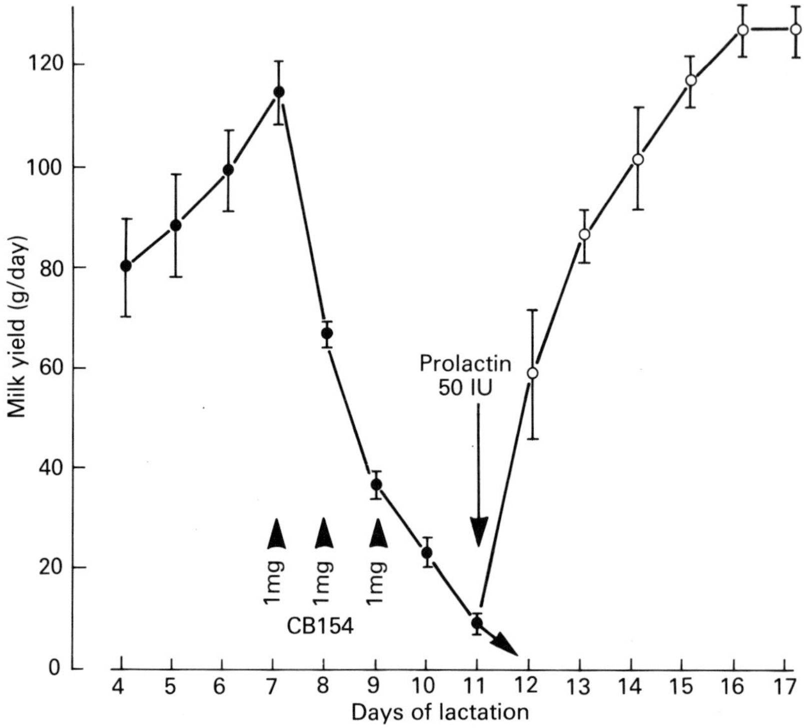

Figure 7.9. Effect of bromocriptine (CB154) on milk yield of five rabbits. On day 11 two of the rabbits were given 50 international units of ovine prolactin (open circles). Bars show standard errors of means.
(Source: Taylor and Peaker, 1975.)

7.9). There is now good evidence for the involvement of GH in galactopoiesis in ruminants. Not only is it possible to substantially increase milk yield by injecting exogenous GH into lactating cows (see section 11.3.3**C**) (and, to a lesser extent, goats and sheep), but blood concentrations of GH are significantly higher in high-yielding than in low-yielding cows. The role of adrenal steroids in established lactation has received rather less attention, but intact adrenal glands would appear to be essential for lactation in both ruminants and non-ruminants. Some studies on lactating goats, in which cortisol concentrations were elevated by injection of a synthetic ACTH, showed significantly increased milk yields, but other studies did not confirm this galactopoietic effect.

Thyroid hormones seem to act in a 'permissive' capacity in galactopoiesis. In both dairy and laboratory animals blood thyroxine concentrations are negatively correlated with milk yield. Somewhat paradoxically, administration of exogenous thyroactive compounds to cows increases yield. Insulin appears to be important in galactopoiesis in non-ruminants, by maintaining both mammary cell numbers and certain enzyme systems. In ruminant species, galactopoietic effects are indirect, and probably exerted via nutrient partitioning (see section 10.4.3).

7.5.1. *Local factors*

We have seen that at lactogenesis chemical factors synthesized in the mammary gland exert a 'local' control over milk secretion. An analogous mechanism is also evident in established lactation. Thus, if one gland of a dairy animal is milked three times daily its yield rapidly increases by comparison with yields of the twice-daily milked glands of the same animal. It has been suggested that the increased yield results from more frequent removal of a chemical inhibitor present in milk, because simply massaging the gland frequently does not elicit the response. There are indications that the inhibitor present in goat's milk is in the whey protein fraction, but its identity has not yet been established. To be effective in enhancing yield by frequent milking, it is insufficient to postulate the mere *presence* of the inhibitor in milk at constant composition; one must envisage that the degree of inhibition increases progressively with time. This could be achieved either by an increasing potency of the chemical (see section 4.9.1) and/or by an increasing sensitivity of the mammary epithelium to its effects: it is possible that the increased sensitivity results from conformational changes in secretory cells consequent on distension of the alveoli with milk.

7.6. Marsupials

A separate section on marsupials is necessary because their mammary development and secretory activity are so different from those of eutherians. Moreover, local factors (see section 7.5.1) exert a very prominent role in these species.

Pregnancy in marsupials is very short and, while the associated mammary growth is dependent on luteal function, the fetus and/or placenta are not involved. After birth, only suckled glands develop, and the remaining glands rapidly regress. Post partum mammary development in marsupials characteristically shows two distinct phases: an early phase of modest growth and a later phase of accelerated and extensive growth. In the tammar wallaby, the first phase lasts about 20 weeks, during which time the milk secreted is low in volume and contains high carbohydrate and low fat concentrations. With the change in growth rate at about 20 weeks, yield and both total solid and fat content increase, while carbohydrate content decreases.

The later phase of lactation is associated with increased blood concentrations of prolactin, which, apart from a brief peak at parturition, had remained at low concentrations during the early phase. However, despite the low blood prolactin concentration in the early phase, its presence seems to be essential for lactation, since suckled glands show a marked increase in prolactin receptors (see section 9.6.1.1) at this time. As will be discussed (in section 9.5.2), lactogenesis in marsupials appears to be unusual in that it does not depend on progesterone withdrawal.

Marsupials usually nurse, concurrently, young which are at quite different stages of development (see section 1.6.2), and it is an indication of the potency of local control mechanisms that the volume and composition of the milk secreted by adjacent glands in the pouch are remarkably distinct, but finely attuned to the needs of the respective young.

Further reading

Anderson, R. R. (1978). Embryonic and fetal development of the mammary apparatus, in *Lactation*, Vol. 4, pp. 3–40, ed. Larson, B. L. (New York and London: Academic Press.)

Ceriani, R. L. (1974). Hormones and other factors controlling growth in the mammary gland: a review. *J. Invest. Dermatol.* **63.** 93–108.

Cowie, A. T., Forsyth, I. A. and Hart, I. C. (1980). *Hormonal Control of Lactation.* (Berlin, Heidelberg and New York: Springer-Verlag.)

Cowie, A. T. and Tindal, J. S. (1971) *The Physiology of Lactation.* (London: Edward Arnold).

Forsyth, I. A. (1983). The endocrinology of lactation, in *Biochemistry of Lactation*, pp. 309–49, ed. Mepham, T. B. (Amsterdam and New York: Elsevier.)

Forsyth, I. A. and Hayden, T. J. (1977). Comparative endocrinology of mammary growth and lactation. *Symp. Zool. Soc. Lond.* **41.** 135–63.

Fulkerson, W. J. (1979). *Hormonal Control of Lactation*, Vol. 1. (St Albans, Vermont: Eden Press, Quebec and Eden Medical Research Inc.)

Josimovich, J. B. and Archer, D. F. (1978). Primate lactogenic hormones and functions, in *Lactation* Vol. 4, pp. 281–312, ed. Larson, B. L. (New York and London: Academic Press.)

Josimovich, J. B., Reynolds, M. and Cobo, E. (1974). (eds) *Lactogenic Hormones, Fetal Nutrition and Lactation.* (New York and London: Wiley.)

Knight, C. H. (1984) Mammary growth and development: strategies of animals and investigators. *Symp. Zool. Soc. Lond.* **51,** 147–70.

Lascelles, A. K. and Lee, C. S. (1978). Involution of the mammary gland, in *Lactation*, Vol. 4, pp. 115–77, ed. Larson, B. L. (New York and London: Academic Press.)

Maule Walker, F. M. (1984). Exocrine and endocrine secretions of the mammary gland: local control in ruminants. *Symp. Zool. Soc. Lond.* **51,** 171–91.

Peaker, M. and Maule Walker, F. M. (1980). Mastectomy and mammary glands in reproductive control in the goat. *Nature (Lond.)* **284,** 165–6.

Raynaud, A. (1971). Foetal development of the mammary gland and hormonal effects on its morphogenesis, in *Lactation*. pp. 3–29, ed. Falconer, I. R. (London: Butterworths.)

Tindale-Biscoe, C. H., Stewart, F. and Hinds, L. A. (1984). Some factors in the initiation and control of lactation in the tammar wallaby. *Symp. Zool. Soc. Lond.* **51,** 389–401.

8 Neuroendocrine Control of Lactation

8.1. Introduction

In most species, over 80% of the secreted milk is located, just prior to suckling, in the lumina of the alveoli and fine-bore ducts (see section 2.2.3), so that active expulsion is required to overcome the resistance to milk flow afforded by such small apertures. In some dairy ruminants, in which a much greater proportion of the milk is accommodated in large ducts and cisterns (perhaps over 50%), the ejection process may not be so crucial in effecting milk removal, but it is likely to be necessary for continuing milk secretion. Removal of milk (by suckling or milking) from the larger ducts of the gland (where it is present as a result of milk ejection or drainage from the alveoli) is regarded as a 'passive' process.

The suckling stimulus is important in three ways, which might be termed the *short-term*, *medium-term* (operative over a period of hours) and *long-term* (operative over hours to days, depending on species) effects. In the short term, suckling-induced ejection allows the young to obtain almost all the milk present in the glands, though a small proportion, the *residual* milk (perhaps 5–10% in dairy cows), is retained in the glands. In the medium term, removal of milk from the alveolar lumina prevents the local accumulation of chemical factors which are believed to inhibit further secretion (see section 7.5.1). In the longer term, the suckling process elicits release into the maternal blood of factors on which the maintenance of the secretory process depends, that is, the galactopoietic hormones (see section 7.5).

8.2. Milk ejection

8.2.1. Definition of the process

Ejection of milk is accomplished by contraction of the myoepithelial cells surrounding each alveolus and longitudinally orientated in the walls of the fine-bore ducts (see Figure 2.3). The ejection process results in the rapid transfer of milk from the lubuloalveolar spaces into larger ducts. It can be readily detected as an increased turgidity of the gland, or, more quantitively, by measuring the increase in intramammary pressure recorded via a catheter inserted into a teat canal. The flow of milk due to passive withdrawal, which is greatly increased following ejection, has been given the name *let down* by dairymen, and *the draught* by medical clinicians. On occasion, the intramammary pressure is sufficiently great to overcome the resistance of the teat sphincter in an unsuckled gland, so that, for example, in cows milk may leak from the teats. When it happens in nursing women it can be inconvenient, but the problem does not usually persist after the first few days of breastfeeding.

8.2.2. The suckling process

Efficient removal of milk from the large ducts requires effort on the part of the young. Erection of the teat (see section 2.2.7) aids its retention in the mouth but contrary to popular belief (encouraged by the term 'suckling') the mouthing action is not primarily one involving suction. The process is mechanically akin to hand-milking a dairy animal, in that the milk in the teat sinus is expelled through the orifice following compression of the base of the teat between the upper gum and the tongue resting on the lower gum; when this compression is temporarily relaxed more milk flows from the gland into the teat sinus as a result of the ejection-induced pressure. In nursing women, the nipple is sucked to the back of the baby's mouth, so that, effectively, a 'teat' is formed, but milk removal is, again, due to compression between the tongue and hard palate.

Very little is known about the act of suckling in other than laboratory and domesticated species. Interspecies variations in suckling frequency are great (see section 3.9.7) and some species, such as the whale and dolphin, have to circumvent formidable problems, since suckling occurs under water, between breaths. In monotremes, in which there are no teats, milk intake involves a lapping or sucking process, and sucking has also been postulated to have been an important behavioural feature of the first mammals in their evolution from reptilian ancestors (see section 1.4).

The act of suckling clearly requires cooperation between mother and neonate, and while hunger might well represent the dominant motivation for the young, the mother's motivation is less readily explicable, though often hardly less pronounced.

In some laboratory animal species there is evidence that maternal behaviour has an endocrine basis; for example, in rabbits the characteristic nest-building activity seems to be triggered by changes in secretion of ovarian steroids and prolactin, but for other species the evidence is more equivocal. In rats, prolactin has been shown to exert an interesting effect on the litter by its control of the production of a pheromone which is excreted in the maternal faeces. The pheromone attracts the young, who are thus able to locate the mother for periods of suckling. However, in species with precocious young vocalization seems to be a common means employed for orientating the young to the mother.

8.2.3. The nature of the milk ejection reflex

The ability of extracts of posterior pituitary gland to evoke a flow of milk from the cannulated teats of lactating goats was first reported by Ott and Scott in 1910, but the physiological significance of the observation was not appreciated: the response was believed to be a pharmacologically induced galactopoietic effect. For many years milk ejection was thought to be a purely nervous reflex, in which intramammary pressure increased as a result of smooth muscle contraction or, according to one theory, engorgement of mammary blood vessels, as occurs in erection of the penis. It was only in the 1940s that the neuroendocrine nature of the reflex was first appreciated.

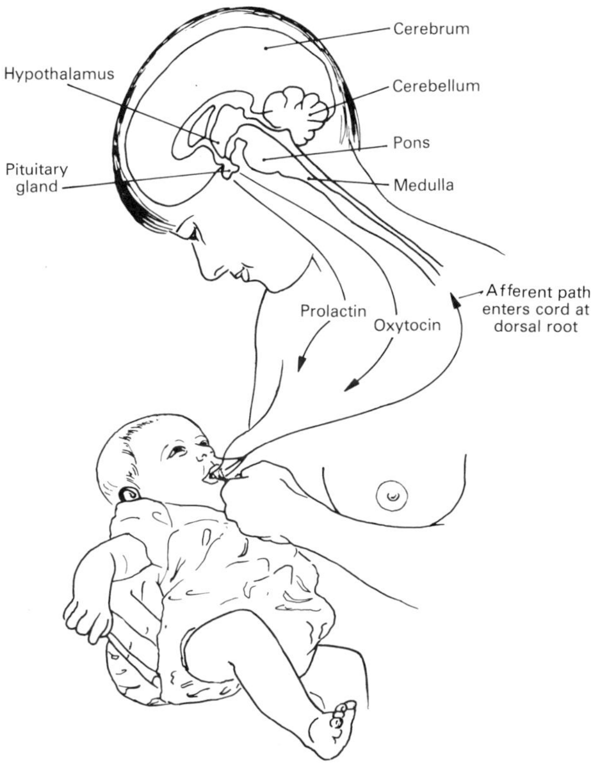

Figure 8.1. Milk ejection in a breastfeeding woman. A simplified scheme indicating organization of the central nervous system.

The reflex is illustrated, in outline, in Figure 8.1. Nerve impulses generated by the suckling young are conveyed via the spinal cord to the posterior lobe of the pituitary gland, at the base of the brain, from where the peptide hormone, oxytocin, is released into the blood. On reaching the mammary gland, by way of the mammary artery, oxytocin stimulates myoepithelial cell contraction.

8.2.4. The afferent path

In describing the components of the reflex we shall follow the advice of the King of Hearts in *Alice in Wonderland* to 'begin at the beginning and go on till you come to the end; then stop'. The teat receptors consist of the unmyelinated nerve endings previously described (in section 2.2.7). Microelectrode recording shows that they are highly activated by the suckling stimulus, and they acquire a greatly increased sensitivity to tactile stimulation in the period immediately following parturition.

Impulses thus generated are conveyed to the spinal cord (see Figure 2.6), entering by the dorsal spinal roots at the level appropriate to the gland's location, and transmitted thereafter in the ascending nerve tracts. Tracing the neuronal pathways has proved to be a difficult task: finding a needle in a haystack is probably rather simpler! Investigations have employed techniques such as spinal cord sectioning, localized ablation and electrical stimulation, all in anaesthetized animals. Most studies have been limited to a handful of species, but even so, marked interspecies variations in the nature of the pathway have been revealed. For example, in the rabbit, the path lies in the ipsilateral (same side as the gland concerned) *ventrolateral column*, whereas in ewes there are both ipsilateral and contralateral paths. Ascending fibres pass through the *medulla* and *pons* regions of the hind-brain and then to the *mesencephalon* (mid-brain) and *diencephalon*, where they bifurcate into two routes—a minor dorsal route, close to the *third ventricle*, and the major ventral route, which passes through the *subthalamus*. The two paths rejoin and terminate in the *hypothalamus*.

Such precise anatomical studies, together with evidence of a more physiological nature, indicate that the afferent path is part of the *spinothalamic system*; in ruminants some fibres are also part of the *spinoreticular system*. The spinothalamic system is activated by stimuli of an abrupt or altering nature, in contrast to the *medial lemniscal* system, which responds to weak stimuli and is capable of acute spatial discrimination. Milk ejection thus does not occur in response to a casual, and perhaps incidental, tactile stimulus, but only to the vigorous and sustained act of suckling.

8.2.5. The oxytocin pathway

The afferent nervous pathway synapses in the hypothalamus with the *magnocellular neurosecretory nuclei*, groups of neurones which synthesize the peptide hormones, oxytocin and vasopressin (ADH). In the guinea-pig, oxytocin is largely synthesized in the *paraventricular nucleus* (PVN), but in other species it is also synthesized in the *supraoptic nucleus* (SON), so that as a generalization both nuclei should be regarded as implicated. (It should be noted that although reference is made to *the* PVN and SON, both of these nuclei are paired, with one each side of the midline.)

The axons of the PVN and SON pass down through the basal hypothalamus and pituitary stalk to enter the posterior pituitary gland (the *neurohypophysis*), a distance of about 4 mm in rats. Here the axons terminate in numerous dilatations

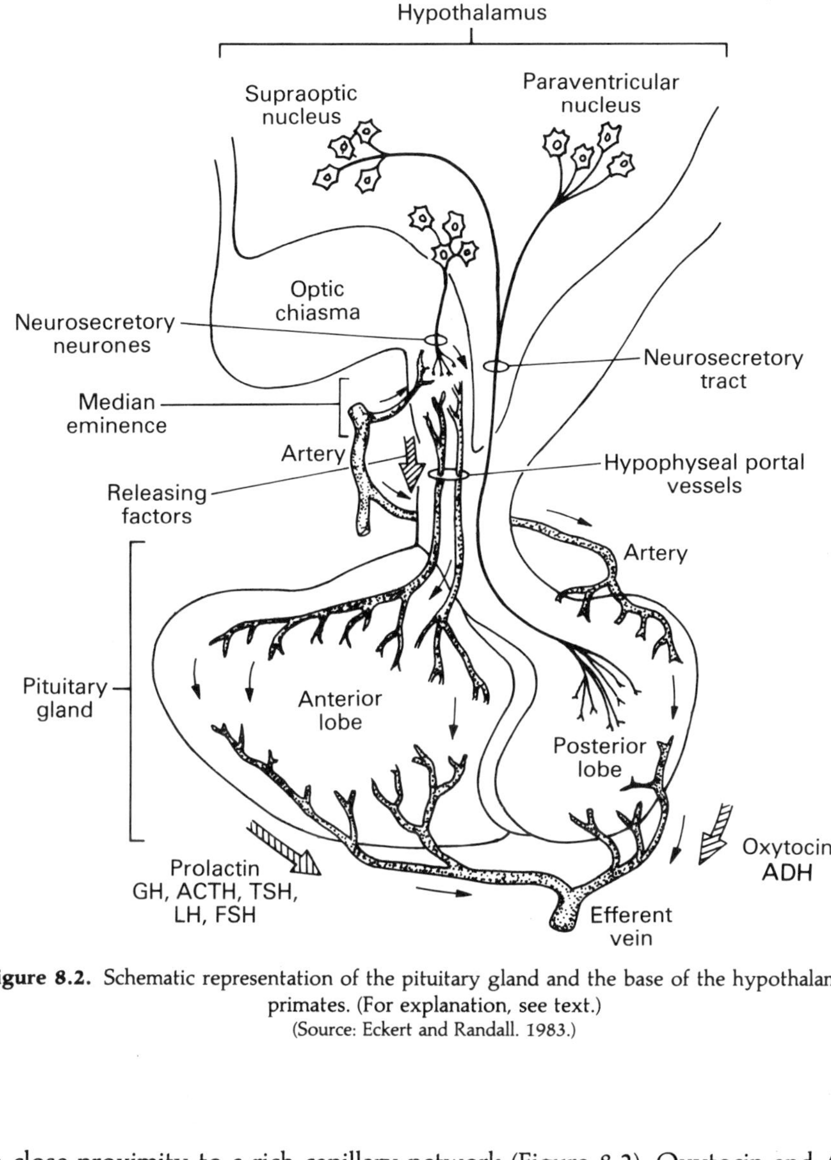

Figure 8.2. Schematic representation of the pituitary gland and the base of the hypothalamus in primates. (For explanation, see text.)
(Source: Eckert and Randall. 1983.)

in close proximity to a rich capillary network (Figure 8.2). Oxytocin and ADH are both synthesized in the form of *prohormones,* that is, glycosylated peptides with molecular weights of about 20 000, of which some components are called *neurophysins*. Following synthesis in the ER of the cell body, and 'packaging' into granules in the Golgi apparatus, the prohormones are rapidly transported along the axons to the pituitary and undergo enzymatic cleavage *in transit,* releasing the active hormone. Consequently, the axonal dilatations normally contain an appreciable store of active hormones available for release into the blood capillaries when the appropriate stimulus arrives. This stimulus consists of the propagated action potentials, initiated by suckling, which on reaching the terminal axonal dilatations elicit discharge of oxytocin granules by exocytosis.

The effectiveness of the physiological stimulus of suckling, as opposed to that due to electrical stimulation, has been thoroughly investigated in rats. Lactating rats only exhibit milk ejection when asleep and it is fortunate that in this (but not in several other) species, most general anaesthetics mimic this situation in a way which does not impair the response. (More recently, it has been possible to confirm several of the observations to be discussed by experiments with unstressed, unanaesthetized rats.) When a litter of hungry pups are allowed to suckle their anaesthetized mother, milk ejection only occurs at roughly 10 min intervals. It is preceded by a burst of electrical activity, recorded by a microelectrode inserted in an 'oxytocin' neurone (see Figure 8.3), and signalled externally by a dramatic response on the part of the pups—they stand up and suckle vigorously. The ejection response, measured with a pressure transducer connected to an intraductal catheter, occurs 10–12 s after the neuronal response, and it has proved possible to break down this interval into three phases: a 2 s period for release of oxytocin into the blood capillaries; 4 s for the oxytocin to reach the mammary glands by way of the circulatory system; and 6 s for activation of the myoepithelial cells.

Microelectrode recording from all four magnocellular nuclei shows that the abrupt bursts of activity occur synchronously in all 'oxytocin' neurones, but activity in the 'ADH' neurones, which is also phasic, appears to be unrelated.

8.2.6. Central control of oxytocin release

Thus, although rat pups in the first 2 weeks of life spend most of their time at the nipple, they are only rewarded with milk at 5–10 min intervals. Milk ejection is, then, ultimately determined by factors arising, consciously or unconsciously, in the mother. There is evidence that in rats, opiates, such as *morphine*, and opioid peptides, such as *β-endorphin* and *enkephalins* (see Table 4.2), reduce the amount of oxytocin released by electrical stimulation. Quite recently, the detection of enkephalins in the oxytocin-secreting neurones has suggested the possibility that direct feedback on the nerve terminals could limit the duration of oxytocin release in a pulsatile discharge phase. In any event, a central 'gating' mechanism for oxytocin release would seem to exist.

It is well known (for example, by nursing women and dairymen) that stress, discomfort, or even unfamiliar surroundings, inhibit milk ejection. The points along the reflex path at which inhibition might occur are numerous. Pain stimuli can block passage of impulses at the spinal level, while fear and apprehension probably act at the hypothalamic level. Since oxytocin reaches the mammary gland via the blood, factors which reduce MBF might also prevent ejection. Thus, stress induces release from the adrenal medulla of adrenaline, a potent vasoconstrictor of mammary vessels.

Central influences on milk ejection may also, however, be facilitatory. As for salivation in Pavlov's dogs, milk ejection may readily become a *conditioned reflex*, elicited in response to the *exteroceptive* stimuli (sights, sounds, smells, etc.)

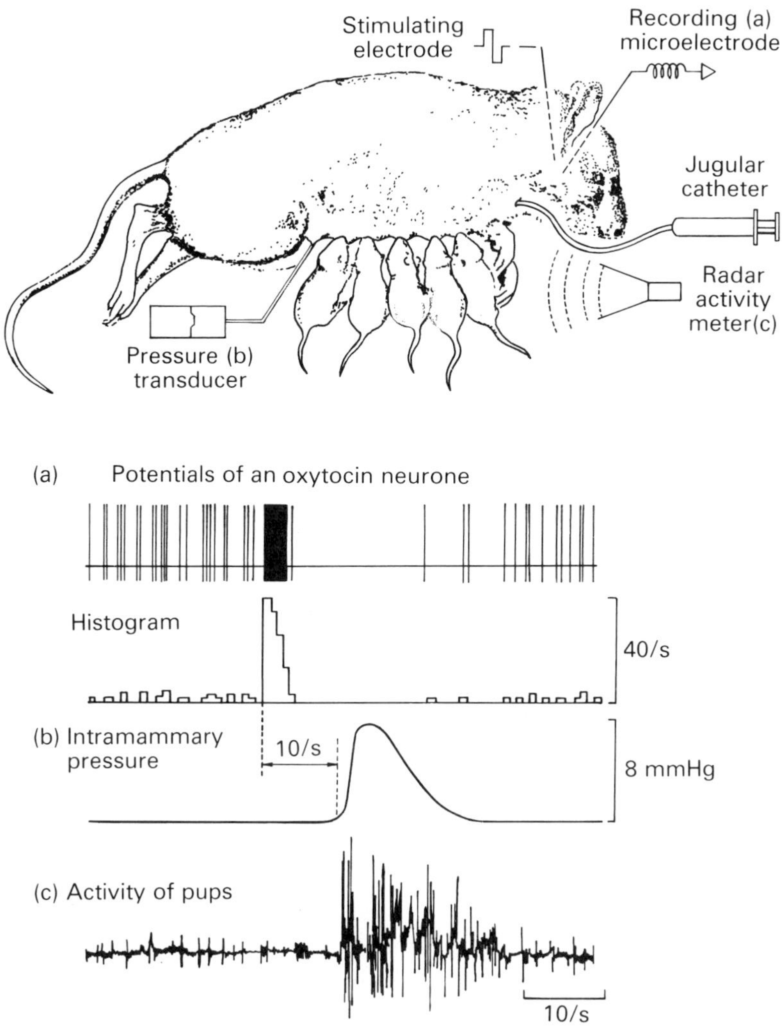

Figure 8.3. Studies on the milk ejection reflex in an anaesthetized rat. (*a*) The explosive increase in number of action potentials in an 'oxytocin' neurone of the PVN induced by suckling. (*b*) The simultaneously recorded changes in intramammary pressure show a peak about 10 s after the major electrical activity. (*c*) Radar recording of pup activity during suckling. (Reproduced from Lincoln 1984, by permission of Cambridge University Press, C. R. Austin and D. W. Lincoln.)

associated with the suckling or milking routine. This explains the fact that, in dairy herds, animals about to be milked show evidence of premature ejection (by milk leakage from the teats, see section 8.2.1) before the milking process begins. Confirmatory, electrophysiological, evidence for a pathway descending from the cerebral cortex to the hypothalamus has been reported.

Milk ejection may also result from vaginal stimulation, a fact which has been appreciated for millenia and led to the age-old practice of blowing air into the vagina of milch-animals to encourage let-down.

8.2.7. *Oxytocin concentrations in blood*

Direct evidence for oxytocin release at suckling is provided by determination of its concentration in blood. This was formerly performed by a bioassay procedure based on intramammary pressure changes in anaesthetized guinea-pigs, but nowadays radioimmunoassay is usually employed.

In all species the release of oxytocin is pulsatile and follows an 'all or nothing' principle—the same amount of oxytocin is released per pulse, producing approximately the same pressure change on all occasions. Oxytocin 'peaks' in the blood are very sharp, with a half-life of about 90 s, but there are marked interspecies differences in the pattern of oxytocin release. For example, in rats, ruminants and human females, there is a series of ejections, separated by intervals of 2–10 min, whereas in pigs there is only a single, brief ejection phase during suckling.

In dairy ruminants it is frequently not possible to detect any oxytocin release in association with milking, yet milk yields in such cases (over 30% in one study on cows) do not differ significantly from occasions when ejection *does* occur. This observation has led to the suggestion that in animals with capacious duct systems milk ejection may be less important than in other species.

8.2.8. *Other ejection stimuli*

There are, however, other factors which need to be considered before we discount the necessity for milk ejection in such animals. We have noted (see section 8.2.6) that ejection occurs in response to conditioned stimuli, so that the transfer of milk from alveoli to ducts may have occurred *before* samples were taken for oxytocin assay in the above studies.

A second possibility is suggested by the observation that plasma kinins, such as *bradykinin*, which are products of the enzymatic proteolysis of plasma proteins, are potent activators of myoepithelial cell contraction. In sheep and goats, bradykinin is particularly effective, but in cows *kallikrein* is more so. The significance of such observations merits further study.

A third factor effecting milk ejection is the, so called, *tap reflex*. As first described in rabbits, if an abrupt tap is applied to the skin over a gland, the myoepithelial cells respond by contracting. Nerves do not seem to be involved because neither local or general anaesthesia nor gland denervation affects the response. The possibility that this is an important mechanism in ruminants is suggested by the vigour with which, for example, lambs are seen to butt the ewe's udder when suckling.

Finally, the sparse motor-innervation of ducts (see section 2.2.6) may have some role to play in the transfer of milk from the lobuloalveolar to the ductal compartments.

8.2.9. *Myoepithelial cell contraction*

The structure of myoepithelial cells has already been described (see section 2.3.6). Coordinated contraction of cells in a single alveolus seems likely to be due to the spread of electrical charge via gap junctions between the cells. Oxytocin diffuses from the alveolar capillaries, binds to specific receptor sites and effects the cell depolarization on which contraction depends. The effectiveness of myoepithelial cell contraction is graphically illustrated by the late J. L. Linzell's drawing of the cells (Figure 8.4).

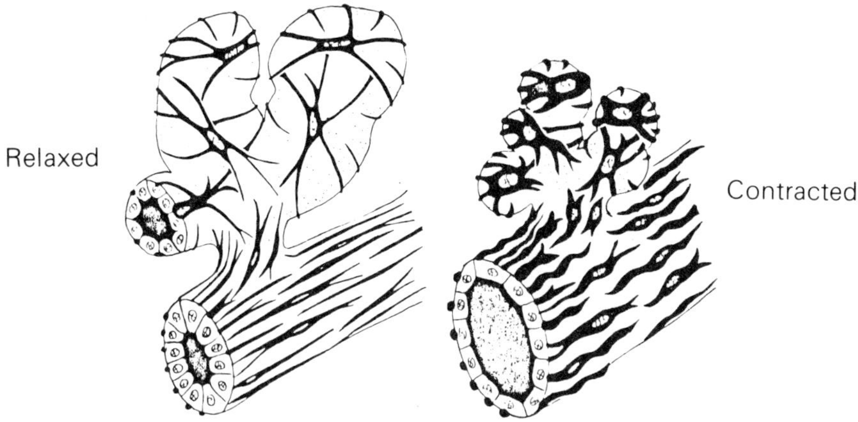

Figure 8.4. The action of myoepithelial cells in contracting mammary alveoli and dilating ducts. (Drawing by the late Dr J. L. Linzell.)

It is possible that myoepithelial cell sensitivity to oxytocin decreases with advancing lactation, and thus leads to gradual involution through less efficient milk removal. Such a decline in sensitivity has been described for marsupials, where it assumes importance in relation to the concurrent feeding of young of quite different stages of maturity (see section 7.6). The myoepithelial cells of the gland suckled by the newborn appear to be sensitive to basal concentrations of oxytocin in blood, whereas those of the gland suckled by the young 'at heel' only respond to the much higher oxytocin concentrations resulting from vigorous suckling.

8.2.10. Summary

The neuroendocrine nature of the milk ejection reflex, in which the efferent path involves the circulatory system, accounts both for the characteristic delay between application of the stimulus and the rise in intramammary pressure (a delay which is, however, often not apparent in 'conditioned' animals) and for the occurrence of ejection in all glands when only one is being suckled. But delay may also be due to the nature of the afferent (neural) component. As it utilizes the spinothalamic system, a sustained, positive stimulus is required, which is highly susceptible to inhibitory influences. A good analogy would seem to be the micturition reflex, which also involves a complex afferent input and has become associated with a range of characteristic species-specific rituals.

8.3. Control of galactopoietic hormone release

8.3.1. The role of the anterior pituitary gland

The suckling stimulus and the exteroceptive stimuli which become effective in the associated conditioned reflexes are also important in controlling release of the galactopoietic hormones previously discussed (see section 7.5, Figure 8.1). The anterior pituitary gland (*adenohypophysis*) is of paramount importance in this respect because the hormones it secretes either have a direct action on mammary tissue or affect it indirectly via release of other hormones, or by causing changes in the metabolic status of the animal.

The anterior pituitary secretes a large number of trophic hormones (see Table 4.1), which, unlike oxytocin and ADH, are synthesized by specific cells in the gland itself. Release of these hormones into the blood is elicited by chemical factors which are synthesized in hypothalamic neurones and reach the anterior pituitary via a system of blood capillaries running in the pituitary stalk, the *hypophyseal portal system* (see Figure 8.2).

Secretion of most adenohypophysial hormones, such as FSH and LH, is controlled by *releasing hormones*, but in the case of prolactin the influence of a *prolactin inhibiting factor* (PIF) is thought to be dominant. Thus, for prolactin secretion to occur, the normal tonic release of PIF must be suppressed. Quite recently, PIF was identified chemically as the catecholamine, *L-dopamine*, but at the time of writing a new claim has been made, that PIF activity is associated with the carboxy-terminus of human placental gonadotrophin releasing hormone (GnRH) precursor. Whether this GnRH-associated peptide (GAP for short) will prove to be more effective in modulating prolactin secretion than is L dopamine remains to be seen, but it is quite possible that more than one substance exerts PIF activity.

In addition to PIF, it has been claimed that a *prolactin releasing factor* (PRF) may also be involved to a minor degree, effecting the rapid release of prolactin occurring in response to stress. The chemical identity of PRF is uncertain, but the fact that injections of *thyrotrophin releasing hormone* (TRH) elicit release of both thyrotrophin and prolactin has led to the suggestion that TRH and PRF may be one and the same substance. Another recent suggestion is that VIP (see Table 4.2) is a physiological mediator of prolactin release. Despite its name, this compound is also widely distributed in the CNS, including the hypothalamus, and in experiments in which antisera to VIP were administered to lactating rats the suckling-induced release of prolactin was markedly depressed. It thus seems possible that prolactin secretion is controlled by a number of chemical factors, some with stimulatory, and others with inhibitory actions.

8.3.2. Prolactin secretion

When discussing the afferent path of the ejection reflex, we noted that at the level of the diencephalon there is a bifurcation into dorsal and ventral routes (see section 8.2.4). Stimuli destined to control prolactin release appear to utilize only the dorsal path. Subsequently, the 'prolactin-pathway' passes to the medial anterior hypothalamic area (*median eminence*), the site of dopamine synthesis. A pathway descending from higher centres merges with the ascending path in the preoptic area.

There has recently been much speculation that endogenous opioid peptides, such as enkephalins and β-endorphins, may be important in controlling prolactin release. Treatment of rats with such compounds elevates blood concentrations of both prolactin and GH, but in goats, the opioid antagonist, *naloxone*, inhibits the basal secretion, but not the milking-induced release, of prolactin. The situation currently seems quite complex, and not amenable to simple generalization, but it appears likely that opioids often act by inhibition of dopamine (PIF) release. The ergot alkaloid derivative, *bromocriptine*, also acts at the hypothalamic level. By binding to the dopamine receptor and mimicking its action it suppresses prolactin secretion, a property which has led to its widespread use in medicine and as a research tool.

Initially, prolactin could only be measured by bioassay in a system which employed estimation of the extent of hypertrophy of pigeon crop glands (see section 1.2), but since the 1960s much more sensitive radioimmunoassay procedures have been available. Studies on goats showed that milking stimulated a rapid prolactin release, with blood concentrations remaining elevated for about 40 min (Figure 8.5), but no release occurred when a denervated gland was

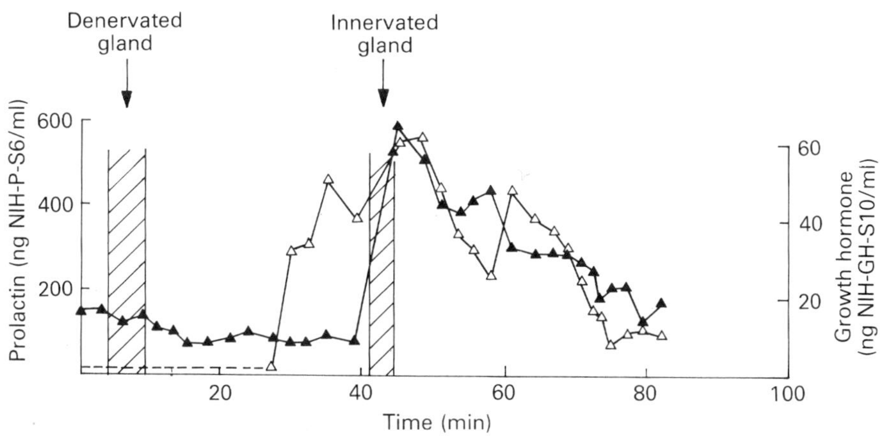

Figure 8.5. Blood plasma concentrations of prolactin (▲) and growth hormone (△) before, during and after milking a denervated and an innervated mammary gland of a single goat. Hatched bar indicates duration of milking.
(Source: Hart and Linzell, 1977.)

milked. Studies on other animals, both ruminant and non-ruminant, confirm the effectiveness of suckling in causing prolactin release.

In the early 1970s, measurement of prolactin release in cows and goats seemed to indicate a positive correlation with milk yield, but subsequently, as noted (see section 7.3.4), seasonal factors were seen to be dominant. Thus, as suggested by experiments involving inhibition of prolactin release (see section 7.5), the presence of *elevated* concentrations of this hormone does not seem to be important in established lactation in ruminants.

In non-ruminant species studied, including human females and laboratory rodents, prolactin release continues to be important throughout lactation. In rats, exteroceptive stimuli become progressively more important in advancing lactation, so that merely the sound or sight of the litter provokes prolactin release.

8.3.3. *Growth hormone secretion*

Release of GH from the pituitary gland at suckling seems to be variable both between and within species, and there is evidence that, at least in some cases, release is not due to stimulation of afferent nerves. For example, Figure 8.5 shows that in a goat, milking of a denervated gland was followed about 20 min later by a pronounced release of GH, although no prolactin release was evident. It is tempting to suggest that GH release may be triggered by metabolic changes consequent on milk removal. On the other hand, it is possible that the variability of response observed may be due to the inability of the radioimmunoassay procedures used to detect the 'secreted' form of this hormone. Other reports suggest that GH release is promoted by opioid peptides (see section 8.3.2).

8.3.4. *ACTH release*

In several species, including rats, sheep and goats, suckling also provokes release of ACTH (see Table 4,1) and, in consequence, secretion of adrenal corticosteroids, but little is known of the neuronal basis of the reflex.

8.3.5. *Pituitary hormone release in non-pregnant animals*

The role of prolactin in the mammogenesis of pregnancy has been emphasized (see section 7.3.4): ovarian steroids appear to play a synergistic, but subordinate role. The fact is underlined by the observation that ovariectomized goatlings can be brought into milk simply by applying the 'milking' stimulus regularly. It is presumed that gland development and milk secretion are initiated by the neurally induced release of prolactin and other pituitary hormones. Lactation can also be induced in non-pregnant women if a child is regularly applied to the breast (see section 11.2.2A).

8.3.6. *Gonadotrophic hormone release*

This is a suitable point at which to mention the inhibitory effect of suckling on pituitary secretion of gonadotrophins (LH and FSH). In some species, such as rats, suppression of release of their hypothalamic releasing hormone, GnRH, produces a condition of *lactational anovulation*. The mechanism of inhibition has not been defined with certainty. We shall return to this topic in Chapter 10.

8.4. Summary

The pathways of oxytocin and prolactin release are for the most part common, with minor deviations occurring in the forebrain (Figure 8.6). Prolactin release

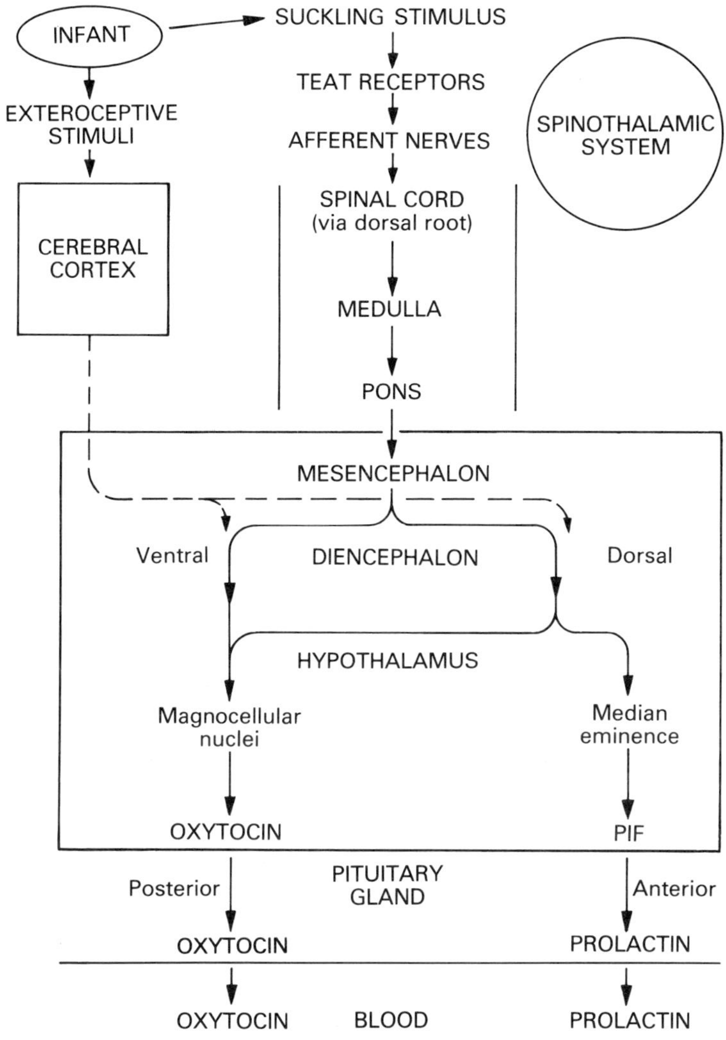

Figure 8.6. A summary of the neural pathways of oxytocin and prolactin release.

seems to be less specifically related to the suckling and associated conditioned stimuli, and environmental factors such as temperature, light and painful stimuli all contribute to maintaining its basal concentration in blood. Its role is not solely related to milk secretion; it is, after all, present in males.

Like GH, prolactin release may be largely sensitive to the metabolic status of the body. It is interesting that blood concentrations of both prolactin and ACTH, which is often regarded as primarily released in response to stressful conditions, are elevated by suckling. Such effects are difficult to understand in view of the inhibitory effect of stress on oxytocin release.

Further reading

Caldeyro-Barcia, R. (1969). Milk ejection in women, in *Lactogenesis*, pp. 229–43, eds Reynolds, M. and Folley, S. J. (Philadelphia: University of Pennsylvania Press.)

Cobo, E. (1974). Neuroendocrine control of milk ejection in women, in *Lactogenic Hormones, Fetal Nutrition and Lactation*, pp. 432–52, eds Josimovich, J. B., Reynolds, M. and Cobo, E. (New York and London: Wiley.)

Cowie, A. T., Forsyth, I. A. and Hart, I. C. (1980). *Hormonal Control of Lactation*. (Berlin, Heidelberg and New York: Springer-Verlag.)

Cowie, A. T. and Tindal, J. S. (1971). *The Physiology of Lactation*. (London: Edward Arnold).

Cross, B. A. (1977). Comparative physiology of milk removal, *Symp. Zool. Soc. Lond.* **41,** 193–210.

Cross, B. A. and Findlay, A. R. L. (1969). Comparative and sensory aspects of milk ejection, in *Lactogenesis*, pp. 245-52, eds Reynolds, M. and Folley, S. J. (Philadelphia: University of Pennsylvania Press.)

Findlay, A. R. L. (1974). The role of suckling in lactation, in *Lactogenic Hormones, Fetal Nutrition and Lactation*, pp. 453–77, eds Josimovich, J. B. and Cobo, E. (New York and London: Wiley.)

Grosvenor, C. E. and Mena, F. (1974). Neural and hormonal control of milk secretion and ejection, in *Lactation*, Vol 1, pp. 227–76, eds Larson, B. L. and Smith, V. R. (New York and London: Academic Press.)

Lincoln, D. W. (1984). The posterior pituitary, in *Hormonal Control of Reproduction*, 2nd Edition (Vol. 3 of *Reproduction in Mammals*), pp. 21–51, eds Austin, C. R. and Short, R. V. (Cambridge and London: Cambridge University Press.)

Shillito Walser, E. (1977). Maternal behaviour in mammals, *Symp. Zool. Soc. Lond.* **41,** 313–51.

Tindal, J. S. (1978). Central pathways in oxytocin and prolactin release, in *Physiology of Mammary Glands*, pp. 305–11, eds, Yokoyama, A., Mizuno, H. and Nagasawa, H. (Tokyo: Japan Scientific Society Press; Baltimore: University Park Press.)

Tindal, J. S. (1978). Neuroendocrine control of lactation, in *Lactation*, Vol. 4, pp. 67-114, ed. Larson, B. L. (New York and London: Academic Press.)

9 Control of Mammary Function at the Cellular Level

9.1. Introduction

In Chapters 7 and 8 we considered the factors believed to be important in regulating mammary growth and function, with emphasis on the changing hormonal 'environment' of the body and on the role of neural stimuli. In this chapter we shall, as it were, 'turn up the magnification' to examine the operation of control processes at the level of the mammary secretory cells, and this will involve drawing together many of the concepts discussed in the preceding four chapters.

9.2. Generalities

Many hormones act by selectively modifying the metabolic activity of particular cell types, by regulating the expression of genes which code for the synthesis of enzymes. Peptide hormones (for example, prolactin, placental lactogen and GH) are believed to bind to specific receptor molecules on the cell surface, and, subsequently, either to enter the cell by endocytosis (*internalization*) or to effect transduction of their 'signal' to the nucleus via *second messengers*, chemicals formed in the cytosol in response to the hormone-binding process. Steroid hormones, by contrast, are believed to bind to intracellular receptors following their entry into the cell: the resulting *hormone-receptor complex* then binds to specific regions of the nuclear chromatin and thereby regulates gene expression.

The receptor molecules, on or in cells, experience *turnover*, so that an important factor governing the ability of cells to respond to hormones is the relative rates of synthesis and degradation of the receptors. Some factors increase receptor numbers (*up-regulation*), others decrease them (*down-regulation*). These factors may be either the hormone specific to the receptors or other hormones or chemicals which do not themselves bind to the receptors. In some cases there may be competition between different hormones for the same receptor.

Regulation of gene expression is not the only way in which hormones control cellular activity. They may, for example, influence the activity of membrane permeability to ions. But because hormones often act on rate-limiting processes, the metabolic response is, typically, greatly amplified in relation to the strength of the initial hormonal signal.

At the simplest level of description, we can divide endocrine effects on mammary tissue into those promoting cell proliferation (*hyperplasia*), through mitogenic activity, and those which enhance the metabolic activity of individual cells. The increased synthetic activity is associated with an increased RNA:DNA ratio and organelle development, that is, *hypertrophy*.

9.3. Mammary cell proliferation

In most species, the major phase of cell-division occurs during pregnancy. An important question concerns the nature of the cells undergoing division: is a certain proportion of *all* cells involved or only a fixed population of *stem cells*? Claims have been made in support of both possibilities, but the evidence of exponential growth rates in late pregnancy (see section 7.2.4) suggests that a proportion of *all* cells is involved. Moreover, since in rats the proportion of dividing cells decreases in late pregnancy there must be a reduction in the cell cycle period to maintain exponential growth.

The mammary fat-pad seems to be essential for normal mammogenesis, providing physical support and local control of the extending ductal and lobuloalveolar systems. Ductal extension in the juvenile is confined to the fat-pad, with cell division being limited to the end-buds of the ducts; however, during pregnancy cells of the ducts themselves appear to be capable of division.

The fact that, in some species at least, cell division occurs during lactation, raises the question of which cells could be involved, because it has long been believed that secretory activity and cell-division are mutually incompatible processes. Cytological differentiation of cells in individual alveoli is apparently an 'all or nothing' phenomenon, so that, if we rule out the concept of a stem cell population, three possibilities are presented: (a) new alveoli proliferate from end-buds; (b) proliferation only occurs in undifferentiated cells of non-secretory alveoli; (c) proliferation occurs in differentiated cells of any alveolus. There is little evidence for possibilities (a) and (b), but recent EM studies suggest that differentiated cells are able to temporarily suspend secretory activity to undergo division; possibility (c) would thus appear to be the preferred hypothesis.

9.4. Endocrine control of hyperplasia

It was noted (in section 7.3.4) that oestrogens, in association with anterior pituitary (and, in some species, placental) and adrenocortical hormones, promote ductal growth, while progesterone is additionally required for lobuloalveolar growth. When we examine evidence for direct effects of hormones on mammary tissue a clearer picture emerges of the 'endocrine hierarchy' and of the levels at which different hormones operate.

9.4.1. In vitro *culture methods*

The methods used most extensively for examining direct effects of hormones involve *in vitro* culture of mammary tissue from laboratory animals. Several approaches have been used. For small mammals (such as mice), *whole glands* may be cultured, but *explants* (small pieces of mammary tissue) have been more widely employed. Both these systems suffer from the disadvantage that a mixture of cell types is present, and to avoid this problem a *primary cell culture* system has recently been introduced. In this case, incubation of mammary tissue with enzymes such as collagenase and hyaluronidase leads to disaggregation and the ability to separate the epithelial cells from stromal cells.

All such preparations remain viable for periods of days to weeks when incubated on appropriate nutrient media under aseptic conditions. It is thus possible to observe the effects of added hormones on cell growth and synthetic activity. It is, however, important to appreciate the unphysiological nature of the preparations. For example, the peptide hormones used have invariably been of large-animal origin (and therefore 'foreign' to the cells cultured) and concentrations of hormones required to elicit responses have often, though not invariably, been unphysiologically high. Moreover, synthetic rates generally fall by over 90% within 24–48 h of establishing a primary culture of lactating cells.

9.4.2. *Responses to insulin, cortisol and prolactin*

In several studies performed in the 1970s using explants of mid-pregnant mouse mammary tissue it was shown that incubation with a combination of insulin, cortisol and prolactin (the *ICP hormone triplet*) led to cell division, cyto-differentiation and, finally, to the elaboration of milk components, over a period of 6 days (Figure 9.1). Insulin was thought to be essential for cell division, cortisol for promotion of differentiation and prolactin for inducing synthesis of milk. More recently, the situation has appeared less clear-cut. For example, it seems likely that insulin is effective by substituting for IGFs (*insulin-like growth factors*, also called *somatomedins*). Moreover, there are clear species differences.

In view of the importance earlier assigned to oestrogen and progesterone in mammary development, it may seem surprising that neither appears essential in culture experiments. In fact, oestrogen alone has no effect on cell growth *in vitro*, and, although it shows some synergy with prolactin, the best evidence is that obtained with cultures of breast cancer cell lines, in which the cytosolic oestrogen receptor has been studied extensively (see section 9.12). Oestrogen may, however, be effective in inducing prolactin receptors on mammary cells. The principal role of oestrogen in mammogenesis would thus seem to be exerted indirectly, by stimulating prolactin release from the anterior pituitary gland, an effect demonstrated in pituitary culture experiments. Little is known about the site of action of the mammogenic effects of progesterone, although, as we shall see, its

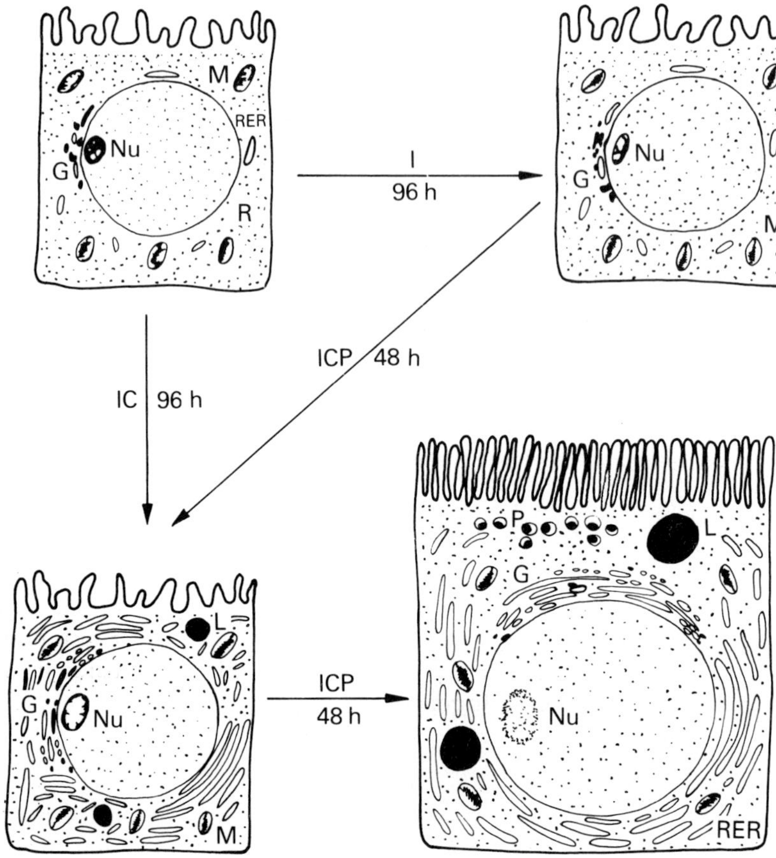

Figure 9.1. Schematic representation of ultrastructural changes in the secretory cells of mid-pregnant mouse mammary explants, when cultured *in vitro* with insulin (I), cortisol (C) and prolactin (P). G, Golgi apparatus; L, lipid droplet; M, mitochondrion; Nu, nucleus; RER, rough endoplasmic reticulum.
(Source: Mills and Topper, 1970.)

inhibitory effect on milk secretion is undoubtedly exerted on mammary cells. There is a similar dearth of information on the direct effects of placental lactogen.

Our knowledge of the hormonal control of hyperplasia is thus rather incomplete. Prolactin (and probably also placental lactogen) seems likely to prove of major importance in stimulating cell division, and while oestrogen may control prolactin secretion, its action, and that of progesterone, at the mammary cell level appears subordinate, and chiefly one of augmenting the effects of the pituitary hormones. Corticosteroids are usually effective mammogenic agents in culture experiments and there is accumulating evidence for the importance of growth factors, synthesis of which may be induced by oestrogens.

9.5. Lactogenic effects

The hormonal control of mammary biochemistry at lactogenesis has received much more attention: brief reference has already been made to early experiments with the ICP triplet in which lactogenesis was induced (see section 9.4.2).

The physiological and biochemical correlates of lactogenesis are myriad; for example, mammary blood flow increases (see Figure 7.6) in association with an increase in cardiac output (see section 10.3), and body metabolism is markedly elevated as the mammary glands' demand for substrates increases. At the ultrastructural level, the mammary secretory cells become highly differentiated, although there are clear differences in the time-course of development of different organelles; for example, in mice, mitochondrial development is complete prepartum, but the ER continues to increase in volume during lactation. The process of cytodifferentiation is most dramatic in laboratory rodents, in which stage II of lactogenesis follows closely on stage I (see section 7.4.1), resulting in cells with the ultrastructure shown in Figure 2.10 (page 24). Ribosomal aggregation into polysomes is rapidly increased at lactogenesis and most ribosomes are bound to ER.

The ultrastructural changes are consequent on activation of the synthetic processes following the 'pulling of the lactogenic trigger'. Our task now is to attempt to define the chain of molecular events which culminates in the full physiological expression of lactogenesis.

9.5.1. The role of prolactin

Early studies with mid-pregnant mouse mammary explants (see section 9.4.2) suggested a crucial lactogenic role for prolactin, since it alone appeared capable of stimulating milk protein and fat synthesis in synergy with other hormones. In some strains of mouse, however, GH was found to be equally as effective as prolactin.

Subsequent studies have emphasized the role of prolactin in stimulating transcription of casein mRNA molecules in cultures of rabbit, rat and mouse explants, increased mRNA accumulation being evident within 60 min of prolactin administration. But induction of casein synthesis is not due solely to stimulation of transcription. There is much evidence that prolactin also (a) stabilizes mRNA against degradation, (b) enhances translation on the ribosomes (partly by increasing rRNA synthesis), and (c) affects certain post-translational events, such as protecting milk proteins from lysosomal degradation and activating the microtubular system. In rat cells, stabilization of mRNA appears to be the quantitatively major effect of prolactin. In most species studied, prolactin has equivalent effects on α-lactalbumin synthesis, although it has been questioned whether this is an absolute requirement in rat cultures.

Prolactin also stimulates mRNA of enzymes involved in the synthesis and secretion of non-protein milk components. For example, it promotes lipoprotein lipase (see section 5.5.1.1) activity in rat mammary tissue and activates membrane-bound $Na^+ - K^+$ activated ATPase (see section 6.5) in rabbit cells. Since α-lactalbumin is the specifier protein (see section 5.4.2) of lactose synthetase, the rate-limiting step in lactose synthesis, prolactin is clearly intimately concerned with control of secretion of the whole aqueous phase of milk, and it seems likely that the detection of lactose in mammary secretions of late pregnant goats is attributable to the analogous effects of placental lactogen, the concentrations of which in maternal blood increase at this time.

The lactogenic effects of prolactin are subject to modulation by other hormones. Cortisol synergizes with prolactin, but in mouse cultures its effects on different proteins are markedly concentration-dependent. Thus, at low cortisol concentrations the synergistic effect on α-lactalbumin synthesis is maximal, but at the higher cortisol concentrations which stimulate casein synthesis, that of α-lactalbumin is inhibited. In rabbit cultures, prolactin is effective in the absence of corticosteroids.

9.5.2. The role of progesterone

While other hormones, such as thyroxine, have been reported to synergize with prolactin, the most effective modulator of prolactin's action is, undoubtedly, progesterone—but in this case all its effects serve to counter the lactogenic process. In cultured mammary cells, prepared from several species, progesterone prevents or reduces the effects of prolactin at virtually all its identified sites of action: transcriptional, translational and post-translational. Figure 9.2 summarizes the influence of prolactin, cortisol and progesterone on milk protein synthesis in mammary cells in culture.

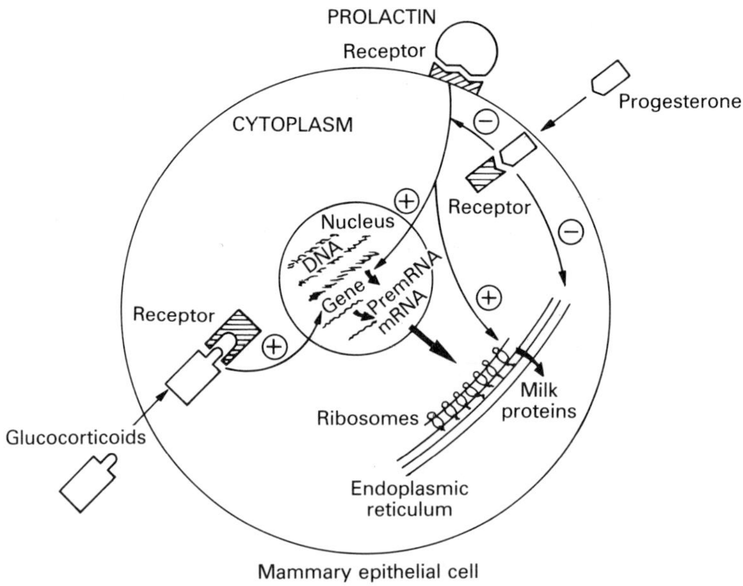

Figure 9.2. Schematic representation of the activation of milk protein genes by prolactin and the effects of its modulators. +, stimulation; −, inhibition.
(Source: Mepham *et al.*, 1982.)

Recent evidence suggests, however, that the control of lactogenesis in marsupials is quite different. Induction of α-lactalbumin synthesis in explants of mammary tissue from tammar wallabies requires prolactin alone; and elevated progesterone concentrations do not inhibit the process. Thus progesterone withdrawal does not seem to constitute the lactogenic trigger in marsupials, even though there is a marked decline in its blood concentration just prior to parturition.

9.6. Transduction of the lactogenic signal

We have seen that prolactin (in the absence of progesterone) is the principal lactogenic agent acting on mammary cells, and that the synergistic action of cortisol is also effective at this site. How is transduction of the signals from the receptor–hormone complexes to the appropriate loci of the nuclear genome achieved?

9.6.1. *Prolactin*

9.6.1.1. Prolactin receptors There is evidence that prolactin not only binds to membrane receptors to exert its lactogenic effects but also, to some extent, becomes internalized. Internalization is not essential for the lactogenic effect in the short-term, but it appears to be important in regulation of prolactin receptor numbers. Thus, if blood prolactin concentrations rise rapidly the internalized prolactin is destroyed in lysosomes, and prolactin receptors are down-regulated. However, when high prolactin concentrations are sustained the receptors are up-regulated, probably because lysosomal degradation is reduced (Figure 9.3). Newly synthesized receptor molecules are transported to the plasmalemma via the Golgi complex (see section 5.3.4).

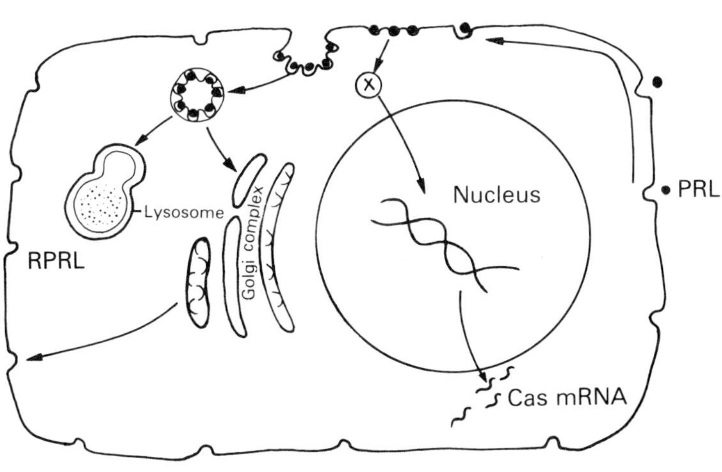

Figure 9.3. Schematic representation of events following the binding of prolactin to its receptors on a mammary secretory cell and the transfer of information to the nuclear compartment. PRL, prolactin molecules; RPRL, prolactin receptors; X, prolactin second messenger.
(Source: Kelly *et al.*, 1984.)

The lactogenic effects are thought to be mediated by a second messenger (X in Figure 9.3) following prolactin's binding to the membrane receptor, but its identity has so far proved elusive. Some of the proposals made are now considered.

9.6.1.2. Cyclic nucleotides In several tissues, *cyclic AMP* (cAMP) is known to act as a second messenger for peptide hormones; binding of the hormone to the receptor activates *adenylate cyclase,* leading to elevation of cytosolic cAMP concentrations. This process does not, however, appear to be involved in prolactin's lactogenic effects. Indeed, cAMP concentrations, which are high in mammary cells during pregnancy, *decline* rapidly at parturition, and in mammary culture experiments cAMP inhibits DNA, RNA and fatty acid synthesis. By contrast, *cyclic GMP* (cGMP) shows reciprocal changes in concentration, increasing at parturition. Moreover, when administered with other agents discussed below, it enhances casein synthesis in explant experiments.

9.6.1.3. Prostaglandins Prostaglandins (PGs) have also been claimed to be mediators of prolactin action. Synthesized from *arachidonic acid,* they have been shown to stimulate RNA synthesis, and, in combination with spermidine (see section 9.6.2), phosphorus incorporation into casein in mouse explants. In mouse, but not in rabbit, culture systems inhibitors of PG synthesis block prolactin's lactogenic effects.

The role of PGs is, however, by no means clear. As noted (see section 7.4.4) $PGF_{2\alpha}$ has been claimed to *inhibit* milk secretion in goats at lactogenesis. Are the explant results valid; do they cast doubt on the inhibitory action claimed for $PGF_{2\alpha}$, or are there marked species differences? Clarification is needed.

9.6.1.4. Peptide messenger In 1984 evidence was presented that in rabbit mammary tissue, the lactogenic effects of prolactin were mediated by a heat-stable, soluble peptide, located in the cell membrane, from which it could be released by the action of proteases.

However, subsequently, workers in the same laboratory admitted that they were unable to confirm these observations. Thus the question of the identity of prolactin's second messenger(s) seems to be back in the 'melting-pot'.

9.6.2. Cortisol

Concentrations of *spermidine* and *spermine* are elevated in many eukaryotic cells undergoing division and differentiation. These polyamines also appear to be intimately associated with lactogenic responses in mammary cells. For example, in rat and mouse cells their concentrations (notably that of spermidine) rise during pregnancy and reach a maximum during lactation. In mid-pregnant mouse mammary cultures, the ICP triplet stimulates spermidine synthesis in advance of protein synthesis. The different hormones of this triplet appear to act at specific points in the sequence of reactions by which polyamines are synthesized from

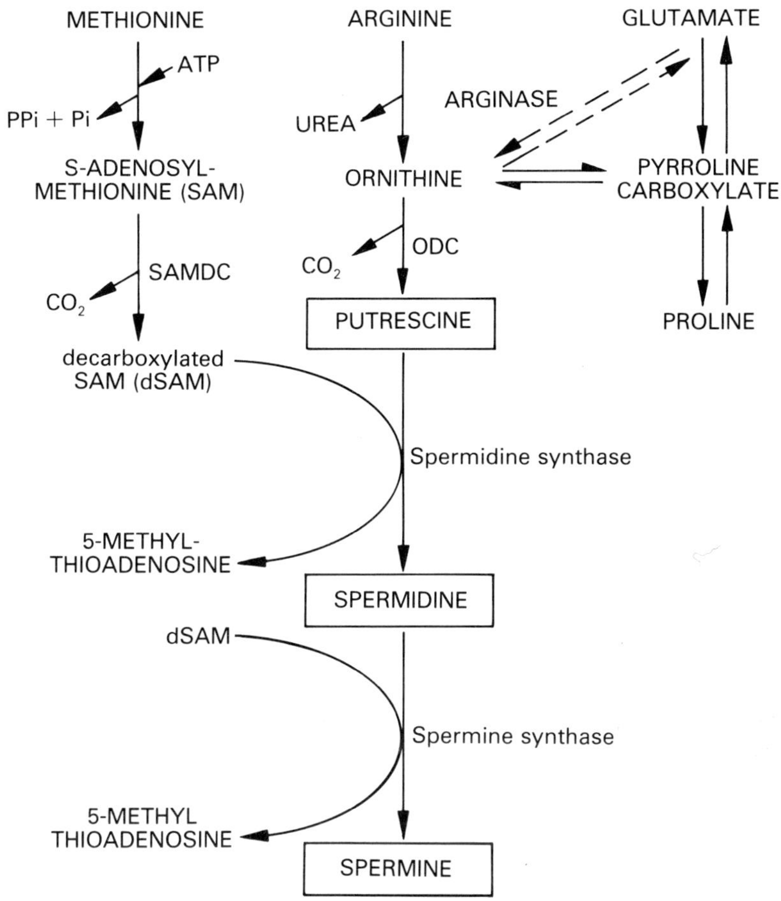

Figure 9.4. Biosynthetic pathways of polyamines in mammalian cells. ODC, ornithine decarboxylase; SAMDC, S-adenosyl methionine decarboxylase.
(Source: Oka, 1983.)

amino acids (Figure 9.4). Thus, the *arginase* and *ODC* stages are stimulated by insulin and prolactin, while *spermidine synthase* and *SAMDC* are stimulated by synergism of insulin and cortisol.

There is strong evidence that, in mice, spermidine is a mediator of the lactogenic effects of cortisol; at concentrations equivalent to those *in vivo* it can substitute for cortisol in the ICP triplet. Its actions appear to be specifically related to promoting α-lactalbumin synthesis, but far less potent for casein synthesis. There is, however, synergy between spermidine and cGMP in stimulating casein mRNA synthesis.

It must be emphasized that species differences are evident, even at this level. In rat culture experiments, spermidine does not synergize with insulin and prolactin, and in rabbit explants prolactin alone suffices to promote milk protein synthesis.

9.7. Effects at the genomic level

There are thus several putative second messengers of hormone action. We need now to consider the way in which the lactogenic stimuli are effective at the chromosomal level by triggering expression of genes coding for milk protein or for key enzymes in anabolic pathways.

Evidence from studies of genetic variation in bovine milks suggests that genes for α_{s1}-, β-. α_{s2}- and κ-caseins are clustered on a single chromosome, while that for β-lactoglobulin is either at a remote locus or located on a different chromosome. Studies using recombinant DNA clones containing complementary DNA for casein and α-lactalbumin have, more recently, enabled localization of milk protein genes in several species, such as rats, mice and guinea-pigs. But localization of genes on particular chromosomes is only one step in the description of the manner of their expression, and theories as to the processes involved are highly speculative.

Repression of genes may be due to masking by the bound chromosomal proteins, that is, chiefly, histones and acidic nuclear proteins. Mediators may thus promote gene expression by inducing conformational (*allosteric*) effects which make available for transcription previously repressed regions of the genome. Some theories propose that gene expression involves dephosphorylation of nuclear proteins, while others implicate reduction in the extent of methylation of DNA bases.

The processes involved in gene expression appear to be exceedingly complex. They are currently the subject of many studies employing the newly acquired techniques of genetic engineering, and it is likely that progress will be rapid. The current 'in-joke' is to label all earlier biochemical research as 'BC', meaning 'before cloning'!

9.8. Mammary gland 'sensing' of progesterone

Cytosolic progesterone receptors have been reported for mouse mammary cells, their concentration declining during pregnancy and becoming undetectable during lactation. The reduced effectiveness of progesterone in inhibiting milk secretion may thus be due to the decline in both blood concentration and mammary receptor numbers. (It is interesting, however, that the *combination* of oestrogen and progesterone *is* effective in inhibiting established lactation, possibly because oestrogen stimulates a restoration of mammary progesterone receptors.) Presumably, during pregnancy the progesterone–receptor complex enters the nucleus and interferes with the lactogenic stimuli promoting transcription. But progesterone has also been shown to have affinity for corticosteroid receptors, so that its effects might be attributable to inhibition of cortisol uptake. There is also some evidence that progesterone exerts its effects by increasing cAMP synthesis.

Apart from this 'direct' sensing of progesterone, it is possible that mammary activity is indirectly affected by progesterone, that is, secondarily to the hormone's effects on other organs. For example, in rats progesterone has been shown to block the glucocorticoid-dependent stimulation of prolactin receptors in mammary cells. Possibly of equal importance are nutritional effects. In rats, there is a fall in food intake just prior to parturition and a marked increase in early lactation, to reach a value 12 days postpartum which is about four times that at parturition. Progesterone has been reported to stimulate food intake, and, since in rats its plasma concentration falls at parturition it may be in part responsible for the change in appetite evident in the periparturient period. It seems likely that the activities of several enzymes known to be peculiarly sensitive to nutritional status would be influenced by such a change. Certainly, short-term starvation and refeeding of lactating rats leads to dramatic and distinctive effects on the rates of synthesis of different milk constituents; in particular, the rate of milk fat synthesis recovers much more rapidly than lactose synthesis in response to refeeding.

Thus, in summary, it is anticipated that progesterone withdrawal will prove to have widespread effects on mammary tissue, by acting both directly and indirectly.

9.9. Enzyme-activity changes in the secretory cells at lactogenesis

The evidence from *in vitro* culture experiments has served to emphasize the role of progesterone withdrawal in triggering lactogenesis, but our description of the biochemical changes occurring has not yet addressed the question of *timing*. Undoubtedly, when milk secretion is established the increased supply of substrates to the cells and frequent removal of products in the milk will enhance synthetic activity in a manner entirely predictable from established precursor-product relationships. The more interesting question is 'which crucial biochemical event precipitates the whole process?'

It might be surmised that all we need to do is to discover which of the many enzymes of mammary cells shows the first dramatic change in activity; this would, surely, indicate the pivotal change we seek. Many studies have been performed to answer this question, particularly on laboratory rodents. In considering the evidence it is useful to assign enzymes, where possible, to one of three groups, depending on whether the major increase in activity occurs (a) well before parturition, (b) at parturition, or (c) well after parturition.

In mouse mammary glands, several enzymes of the glycogenesis pathway show increased activity prepartum, leading to a surge of glycogen deposition in the cells which rapidly declines coincident with increasing lactose synthesis. It has been reasoned that this represents a mechanism by which adequate glucose can be ensured, should the demands of lactose synthesis exceed glucose supply from the blood at parturition. The prepartum rise in glutamate dehydrogenase activity might satisfy a similar contingency in respect of amino acid supply.

Perhaps the most significant of the enzymes in group (b) is lactose synthetase. In mice, a marked asynchrony is evident in the time-course of changes in the galactosyltransferase and α-lactalbumin subunits (see section 5.4.2) of this enzyme. There is a sharp increase in the concentration of α-lactalbumin at parturition (Figure 9.5), leading to modification of the substrate specificity of galactosyltransferase, the activity of which has been increasing progressively in the

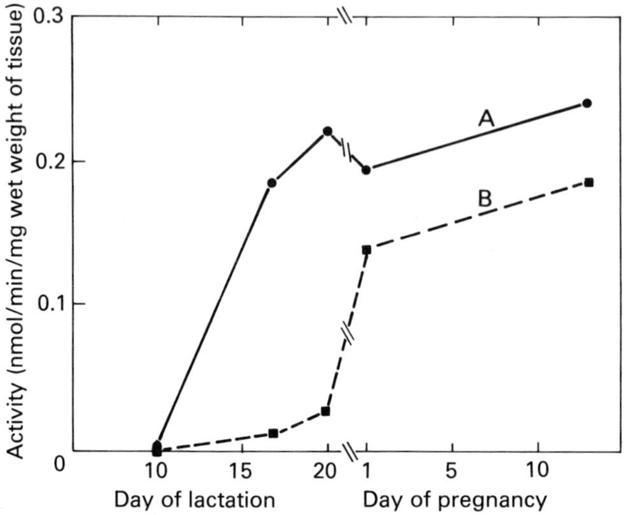

Figure 9.5. Concentrations of galactosyltransferase (A) and α lactalbumin (B) in homogenates of mammary tissue from mice at different stages of pregnancy and lactation. (Source: Turkington *et al.*, 1968.)

latter half of pregnancy. In rats, induction of lactogenesis by ovariectomy (that is, progesterone removal) leads to a four-fold increase in galactosyltransferase activity and a 30-fold increase in particulate α-lactalbumin: both increases are prevented by injection of exogenous progesterone. As noted (in section 9.5.2), progesterone withdrawal also permits a rapid synthesis of casein mRNA, which is, in fact, more rapidly translated than that of α-lactalbumin.

In group (c) are enzymes associated with milk fat synthesis; for example, in rabbits, rats and guinea-pigs, acetyl coA carboxylase, ATP citrate lyase and fatty acid synthetase show increased activity only after parturition. While progesterone withdrawal is necessary for these changes, it is not sufficient: removal of milk from the gland is also required. It has been suggested that in rats cAMP in milk may inhibit enzyme activity, in a similar manner to that proposed for $PGF_{2\alpha}$ in goats (see section 7.4.4).

Glucose-6-phosphate dehydrogenase (G6PD, an enzyme of the pentose phosphate pathway), the activity of which is closely related to mammary lipogenesis (see section 5.6.2), also falls into group (c). Its activity is, however, unaffected by progesterone and largely determined by glucose flux, so that this change may be seen as secondary to the initial lactogenic stimulus.

In general, enzymes not specifically concerned with synthesis of milk components, the so-called *housekeeper* enzymes, are already active prior to parturition, and show predictable changes associated with increased demand for ATP, for example.

9.9.1. The interpretation of enzyme activity changes

Before jumping to conclusions about the site of operation of the lactogenic trigger, some cautionary words are required.

(a) Enzyme activities are commonly assessed *in vitro* under conditions of optimal substrate and co-factor supply, and what is recorded is the V_{max}, the maximal rate of product synthesis. But such conditions may not apply *in vivo*; the distribution of substrates and co-factors between subcellular compartments may also be different. Indeed, in studies on bovine mammary tissue it has proved impossible to detect significant changes in the activity of any enzyme over the *periparturient* period, despite the obviously increased synthetic activity occurring at this time. In fact, the *amount* of an enzyme, as measured by *in vitro* V_{max} studies, is only one of several factors determining its catalytic activity in a tissue, and may not always be the most important.

(b) The relative increases in activity of different enzymes, assessed *in vitro*, do not necessarily indicate relative rates of synthesis because some enzymes in a pathway may have high inherent catalytic power or have been present at high concentration well before parturition.

What is really required, therefore, is a measure of the changes in *flux* along different metabolic pathways, and very few experiments have been performed to obtain such data. N. J. Kuhn and colleagues, on whose analysis the discussion in this section has largely been based, have made measurements of glucose flux in lactating rats *in vivo*, and substrate uptake by AV difference measurements has also been determined in goats at parturition. Figure 9.6 shows the time-course of mammary glucose uptake and blood concentrations of progesterone and prolactin in goats, which suggests a close inverse relation between the former two.

9.9.2. Enzyme changes at lactogenesis: a summary

Clearly, many of the relevant data are unavailable, but a plausible hypothesis for the sequence of lactogenic events in laboratory rodents runs as follows.

Removal of the inhibitory effects of progesterone permits the lactogenic hormones to stimulate genes coding for casein and α-lactalbumin mRNA. Translation of the latter enables completion of lactose synthetase in Golgi vesicles and initiates water flow by osmosis. The increased demand for glucose may be met by glycogenolysis from stores deposited prepartum (as in mice) and/or by increased uptake from the blood. Activation of rate-limiting enzymes in fatty acid synthesis, such as acetyl CoA carboxylase, leads to an increased demand for NADPH (see section 5.6.2), and hence to enhanced activity of the pentose phosphate pathway. Suckling, by effecting the removal from the gland of an inhibitor(s) in milk, permits the induction of other enzymes involved in lipogenesis.

In ruminants, progesterone withdrawal again seems to be the lactogenic trigger, although a low, but gradually increasing, synthetic activity proceeds in the latter

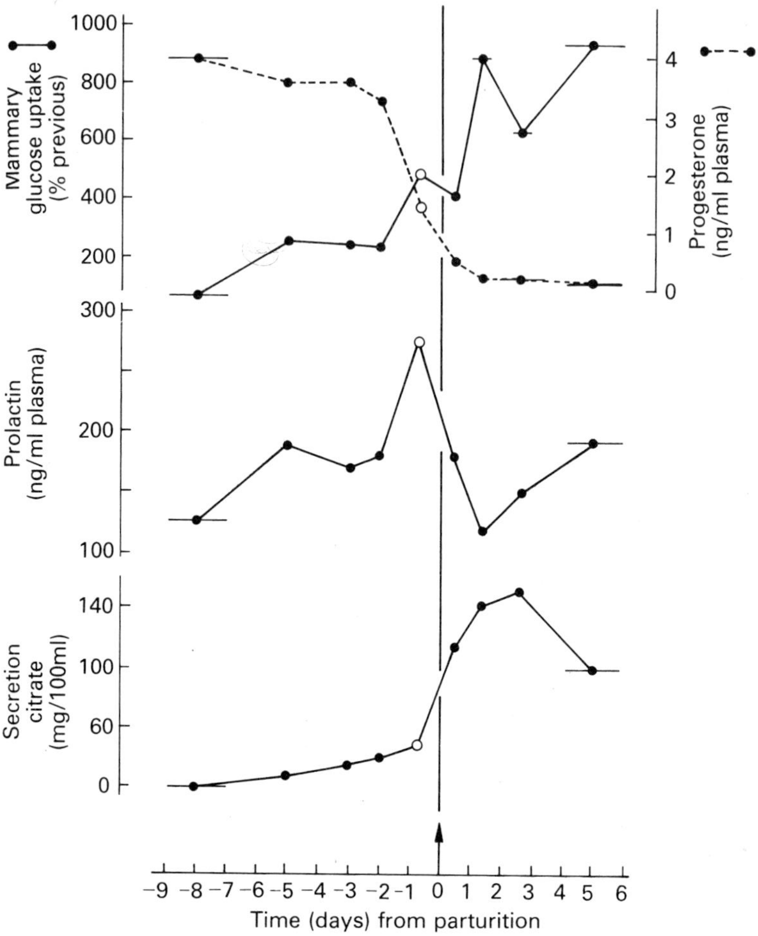

Figure 9.6. Mammary uptake of glucose, arterial plasma progesterone and prolactin concentrations, and concentration of citrate in mammary secretions of six goats. Open circles indicate the first occasion when the value was significantly different from that obtained 7–9 days *prepartum*. (Source: Davis *et al.*, 1979.)

half of pregnancy (stage I), possibly in consequence of high placental lactogen concentrations. Since *in vitro* enzyme assays show little change at parturition, it may be questioned whether the assays are physiologically valid and/or whether synthetic rates are not more closely related to substrate or co-factor supply.

9.10. Cytoskeletal changes at lactogenesis

It was noted earlier (see section 6.2) that the cytoskeleton appears to be important in effecting intracellular transport of milk constitutents. The polymerization of tubulin, that is, microtubule formation (see section 2.3.4) occurs very rapidly at lactogenesis, and it may be surmised that definition of the factor(s) controlling the process may prove crucial to the explanation of the molecular events of lactogenesis as a whole.

We have seen that the lactogenic response which follows ovariectomy in pregnant rats can be prevented by injection of progesterone (see section 7.4.2). Very recently it has been discovered that progesterone injections, but not

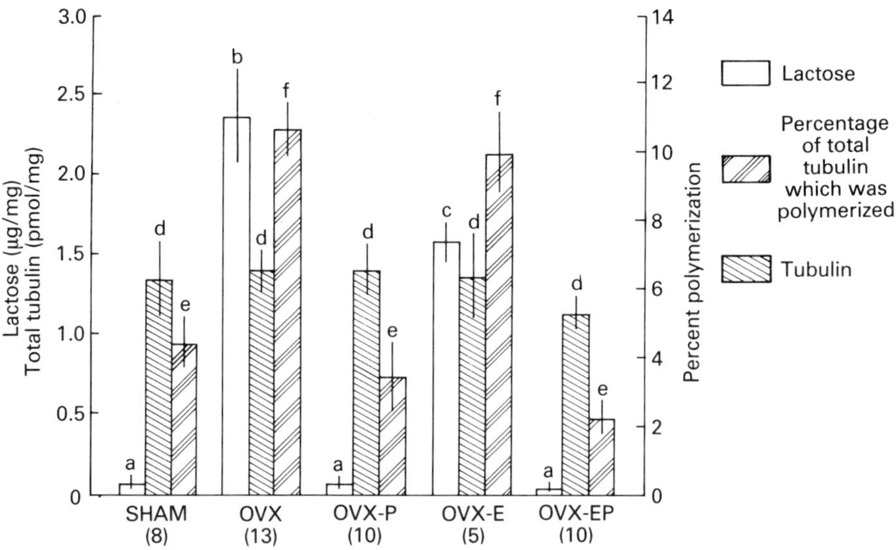

Figure 9.7. Effects of oestrogen (E) and progesterone (P) replacement on the ovariectomy-induced (OVX) increase in mammary gland polymerized tubulin in rats. Bars show standard errors of means; bars with same letters indicate treatment groups which are not significantly different from each other. Number of animals in each condition is shown in brackets.
(Source: Loizzi, 1985.)

oestrogen, also prevent polymerization of tubulin (Figure 9.7). There thus appears to be a close relationship between lactose synthesis and cytoskeletal organization (but not the *amount* of tubulin), both being stimulated by progesterone withdrawal. Little is known with certainty about the mechanisms controlling microtubule formation, but one theory invokes the participation of *microtubule associated proteins*, while others implicate changes in local cytosolic concentrations of Ca^{2+} ions and/or cAMP.

9.11. Maintenance of secretory activity

At first sight, the secretion into milk of α-lactalbumin, a crucial element of the lactose-synthesizing machinery, might appear perverse. On reflection, it provides a simple means of fine control over milk secretion as a whole. The continuous loss of α-lactalbumin from the cell means that lactose synthesis, and consequently, milk flow, will only be sustained as long as the signal for stimulation of the α-lactalbumin gene continues to be applied. Since suckling induces galactopoietic hormone release, the process appears perfectly comprehensible: as long as the demand is registered, the milk will flow.

The situation is not, however, quite so straightforward. For example, in ruminants, although elevated prolactin concentrations are essential to initiate lactation they do not appear to be necessary during established lactation (see section 7.5). The identity of the hormones directly affecting mammary cells during established lactation and their mechanisms of action, have received relatively little study, although currently they are claiming increasing attention.

In vitro culture experiments show that in the absence of the ICP triplet, casein synthesized by rat mammary cells is rapidly degraded by lysosomal activity (see section 4.11). The mammary cells thus seem remarkably parsimonious—any sign that their product is unwanted elicits a prompt curtailment of activity.

9.11.1. Cell activity and cell number

Lactogenesis does not, in itself, result in the full hyperplasic and hypertrophic development of the mammary gland. Yield continues to increase for a period post partum, and this might in part be attributable to increased effectiveness of the endocrine stimuli by means of increased receptor numbers: prolactin has been reported to effect up-regulation of its own receptors in mammary glands of rats and rabbits (see section 9.6.1.1). Injections of prolactin increase yields in laboratory rodents, but the effect is not always solely on synthetic activity. Thus, in rabbits, prolactin increases the K:Na ratio in milk by tightening the 'leaky' tight junctions (see section 6.6), which are a characteristic of this species. There is little evidence for a direct action on mammary cells of either GH or thyroxine, both of which increase yields in lactating ruminants (see section 7.5).

Figure 9.8 shows results of a rare study on the apportionment of changes in milk yield to cell numbers and cell activity, respectively. It is evident that in these rats, 75% of the increased yield in early lactation was attributable to increased DNA_t, while after peak yield the declining activity per cell accounted for all of the reduction in yield. The situation in other species is probably different, since in dairy animals DNA_t declines in late lactation. The factors regulating the persistence of cell activity and cell death are, however, largely undefined.

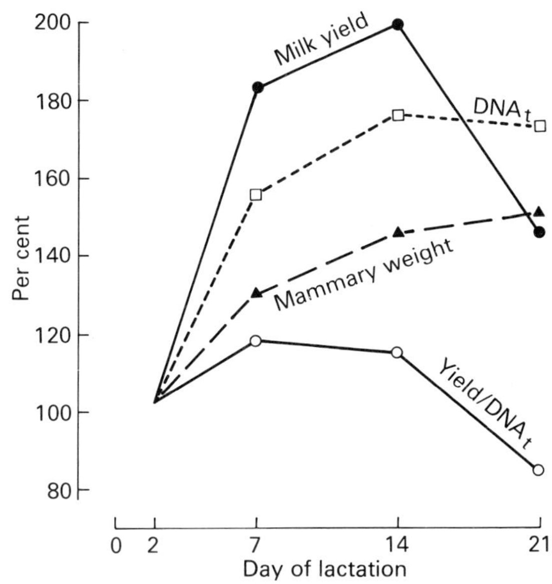

Figure 9.8. Changes in milk yield, mammary weight, DNA_t and yield/DNA_t during lactation in rats. Values are expressed as a percentage of those on day 2 of lactation.
(Source: Knight *et al.*, 1984.)

9.12. Mammary cancer

The susceptibility of mammary tissue to the influence of steroid hormones would appear sometimes to have grievous consequences. There is strong circumstantial evidence that the prevalence of breast cancer is attributable to the regular, periodic exposure to elevated oestrogen concentrations, particularly between the menarche and the first pregnancy. Women whose first pregnancy is before 18 years have only one-third the risk of breast cancer of those having their first child at 35 years. But overall, in both Britain and the United States, breast cancer is the commonest cause of death among women of 35–40 years; and any individual woman has a one in 14 chance of developing the disease during her lifetime.

Breast cancer begins as a local, microscopic *neoplasm,* which by unrestrained cell division produces a clinically recognizable tumour. Cancer cells do not respect the normal 'territorial' rules of growth (see section 9.3), but proliferate uncontrolledly and invade surrounding tissues. Lacking gap junctions (see section 2.3.2), they break away from the initial site and, by entering the lymphatic system, set up *metastases* (secondary deposits), initially in the axillary lymph nodes, and subsequently in other organs of the body.

Only about 50% of patients are cured by surgery and/or radiation therapy of the primary tumour, and the majority of the remainder subsequently develop metastatic disease. Until very recently, the usual treatment in these cases was to administer cytotoxic drugs (that is, *chemotherapy*), which although they affect all cells have their major impact on rapidly dividing cells, including the cancer cells.

But within the last 10 years an effective *adjuvant* (reinforcing) therapy has been introduced based on the endocrine responsiveness of certain tumours.

Normally, oestrogens appear chiefly to have an indirect mammogenic effect, via stimulation of prolactin release (see section 9.4.2); but in mammary cancer cells the presence of oestrogen receptors (ER+) seems to serve as an effective index of the susceptibility of the tumour to hormone therapy, which may indicate that the steroid is a causal factor in the disease. On the assumption that tumour cell progesterone receptors (PR) are induced by oestrogen (see section 9.8), patients in whom both types of receptor have been detected (that is, ER+ and PR+), show a 75–80% responsiveness to endocrine therapy, while in 'ER−, PR−' patients it is only 20–40%. Endocrine therapy may take different forms, such as (a) endocrine gland ablation (ovariectomy, adrenalectomy); (b) administration of high doses of oestrogens (a rather paradoxical but sometimes effective form of treatment); and (c) administration of antioestrogens (such as *nafoxidine*). Use of antioestrogens is now a generally accepted treatment for postmenopausal women, although little is known of the molecular basis of action. Recently it has been proposed that antioestrogens be administered as prophylactic agents to women identified as being 'at risk'.

The role of a mammary tumour virus in breast cancer, described in mice, has not been substantiated for humans.

The suggestion that lactation may reduce the incidence of cancer by a local effect (see section 1.8.1) clearly merits further study.

Further reading

Baum, M. (1981). *Breast Cancer: The Facts.* (Oxford: Oxford University Press.)

Brennan, M. J. (1978). Lactation and the breast cancer process, in *Lactation*, Vol. 4, pp. 313–36, ed. Larson, B. L. (New York and London; Academic Press.)

Cowie, A. T., Forsyth, I. A. and Hart, I. C. (1980). *Hormonal Control of Lactation.* (Berlin, Heidelberg and New York: Springer-Verlag.)

Craig, R. K. and Campbell, P. N. (1978) Molecular aspects of milk protein biosynthesis, in *Lactation*, Vol. 4, pp. 387–404, ed. Larson, B. L. (New York: Academic Press.)

Delouis, C., Djiane, J., Houdebine, L. M. and Terqui, M. (1980). Relation between hormones and mammary gland function. *J. Dairy Sci.* **63**, 1492–513.

D'Orsi, C. J. and Wilson, R. E. (1983). (eds) *Carcinoma of the Breast: Diagnosis and Treatment.* (Boston, Toronto: Little, Brown and Co.)

Houdebine, L. M., Djiane, J., Dusanter-Fourt, I., Martel, P., Kelly, P. A., Devonoy, E. and Servely, J. L. (1985). Hormonal action controlling mammary activity, *J. Dairy Sci.* **68**, 489–500.

Kelly, P. A., Djiane, J. Katoh, M., Ferland, M. H., Houdebine, L. M., Teysott, B., and Dusanter-Fourt, I. (1984) The interaction of prolactin with its receptors in target tissues and its mechanism of action. *Rec. Prog. Horm. Res.* **40**, 399–439.

Knight, C. H. and Peaker, M. (1982). Development of the mammary gland. *J. Reprod. Fert.* **65**, 521–36.

Kuhn, N. J. (1983) The biochemistry of lactogenesis, in *Biochemistry of Lactation*, pp. 351–79, ed. Mepham, T. B. (Amsterdam and New York: Elsevier.)

McGuire. W. L. (1980). Steroid hormone receptors in breast cancer treatment strategy. *Rec. Prog. Horm. Res.* **36**, 135–56.

Mercier, J. C. and Gaye, P. (1983). Milk protein synthesis, in *Biochemistry of Lactation*, pp. 177–27, ed. Mepham, T. B. (Amsterdam and New York: Elsevier.)

Neville, M. C. and Berga, S. (1983). Cellular and molecular aspects of the hormonal control of mammary function, in *Lactation: Physiology, Nutrition and Breast-feeding*, pp. 141–77, eds Neville, M. C. and Neifert, M. R. (New York and London: Plenum.)

Oka, T. (1983) Intracellular mediators of lactogenic hormones, in *Biochemistry of Lactation*, pp. 381–94, ed. Mepham, T. B. (Amsterdam and New York: Elsevier.)

Rosen, J. M., Matusik, R. J., Richards, D. A., Gupta, P. and Rodgers, J. R. (1980). Multihormone regulation of casein gene expression at the transcriptional and posttranscriptional levels in the mammary gland. *Rec. Prog. Horm. Res.* **36,** 157–93.

Short, R. V. and Drife, J. O. (1977) The aetiology of mammary cancer in man and animals. *Symp. Zool. Soc. Lond.* **41,** 211–30.

10 Maternal Physiology During Lactation

10.1 Introduction

While much of the preceding discussion has focused on the activity of the mammary gland, it will have become apparent that lactation makes demands on the body of such a magnitude that the physiology of the mother differs greatly from that prevailing in the non-lactating state. Apart from behavioural changes, we can divide the effects of lactation into three main categories—metabolic, cardiovascular and endocrine—but it should be appreciated that such a division is largely a matter of convention: all these effects are subtly coordinated.

In relating the physiological changes it is, perhaps, even more useful to consider maternal physiology under two main headings: (a) the metabolic consequences, by which it is ensured that the mammary glands are supplied with nutrients adequate to sustain an appropriate level of secretory activity; and (b) consequences for fertility, which in many species is inhibited during lactation, thus ensuring that survival of the suckling young is not compromised by competition for available nutrients with young *in utero*. In a sense, then, the changes are all directed to safeguarding the reproductive investment, and the endocrine system, largely modulated by neural inputs, is a key factor in all aspects of control.

10.2. Metabolic consequences of lactation

Smaller mammals have higher milk yields, greater energy outputs in milk and larger mammary glands per unit of body weight, than do larger mammals (see section 3.8). In very small mammals, lactation imposes an enormous demand on the body's metabolism; for example, the pygmy shrew, weighing 5 g, doubles its food intake when lactating, consuming four times its own weight per day. However, in highly selected dairy animals metabolic demands are also formidable: food consumption increases by more than 50% in high yielding cattle, the daily dry matter intake being about 4% of body weight. At peak lactation a Friesian cow yielding 40 kg of milk per day secretes about 2 kg of lactose and 1.5 kg each of fat and protein daily. The whimsical notion that 'we should regard the cow as an appendage to the udder, rather than the reverse' expresses more than an element of metabolic truth. Figure 10.1 shows typical lactation curves for six species: using the milk composition data given in Chapter 3 a rough idea of the quantities of the major milk components secreted daily by women, cows and rabbits can be calculated.

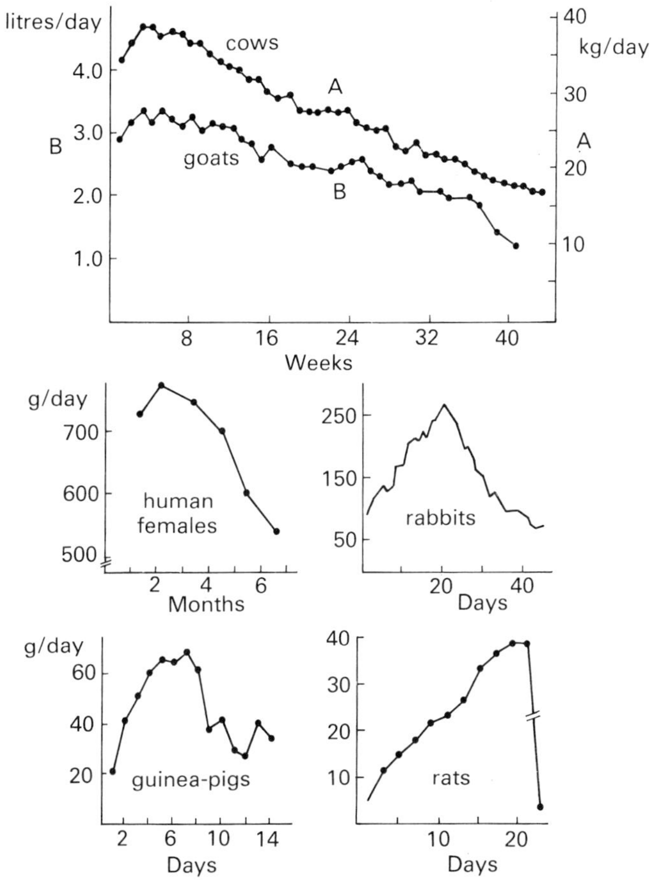

Figure 10.1. Lactation curves of six species, illustrating typical mean yields at different stages of lactation.
(Source: Mepham, 1983.)

Early in the nineteenth century it was widely believed, on the basis of presumed similarities in their chemical compositions, that milk was derived from chyle: there was thought to be a direct lymphatic route from the intestines to the mammary glands, so that the latter had no anabolic role but simply served as a filter. Only much later in the century was this theory found untenable when it became apparent that the milk precursors are drawn from the blood, the same metabolic pool from which all tissues derive their nutrients. The increased nutrient demand of the lactating mammary gland is not, however, met solely by concomitant elevation of food intake. High yielding cows, for example, are in negative energy balance for about 10 weeks post partum (Figure 10.2), with mobilization of body

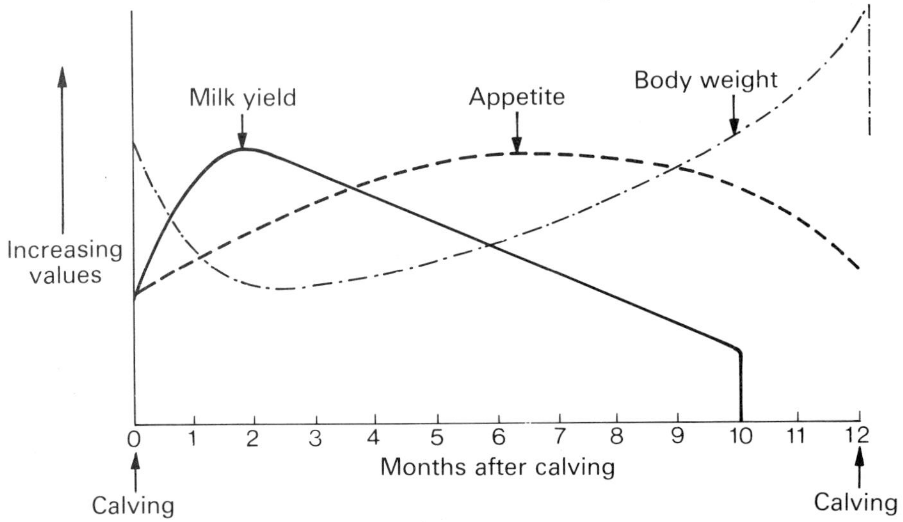

Figure 10.2. Changes in milk yield, body weight and appetite of dairy cows during lactation.

fat being equivalent to about 30% of the energy in the secreted milk. Preparation for lactation thus begins prepartum, and the nutrient pool from which the mammary glands withdraw milk precursors is continuously 'topped-up', both by increased intestinal absorption and by mobilization of body reserves. It has been suggested that the anatomical and biochemical organization of adipose tissue reserves in mammals is finely adapted to the requirement for rapid mobilization post partum.

It might be imagined that the well-known homeostatic mechanisms of the body would guarantee appropriate substrate supply during lactation: the mammary glands would, by dint of their own activity, provide metabolic signals of sufficient amplitude to ensure an adequate *entry rate* (*flux*) of nutrients into the blood and, by the same token, absorb the lion's share of what became available. In fact, the metabolic changes induced by lactation appear to be even more radical. In a very real sense, nutrients are *directed* to the glands and away from certain other tissues. This modified *partitioning* of nutrients is considered to be a consequence of *homeorhetic* control. If homeostasis is likened to the operation of a thermostat, homeorhesis is analogous to changing the temperature-setting of the thermostat; in other words, it describes the coordinated changes associated with the attainment of a new physiological state.

What is the nature of such changes and how are they controlled? It has been appreciated since the 1940s that the endocrine system is of paramount importance, and in recent years the processes involved have been the subject of close scrutiny. Essentially, the hormonal changes associated with lactation appear to regulate the metabolism of body tissues in ways which enhance the entry rate of substrates into the metabolic pool, diminish their utilization by tissues inessential to lactation, augment their partitioning to the mammary glands by cardiovascular control and facilitate their mammary uptake by stimulating the appropriate transport systems located on mammary cells. The sites of control are summarized schematically in Figure 10.3.

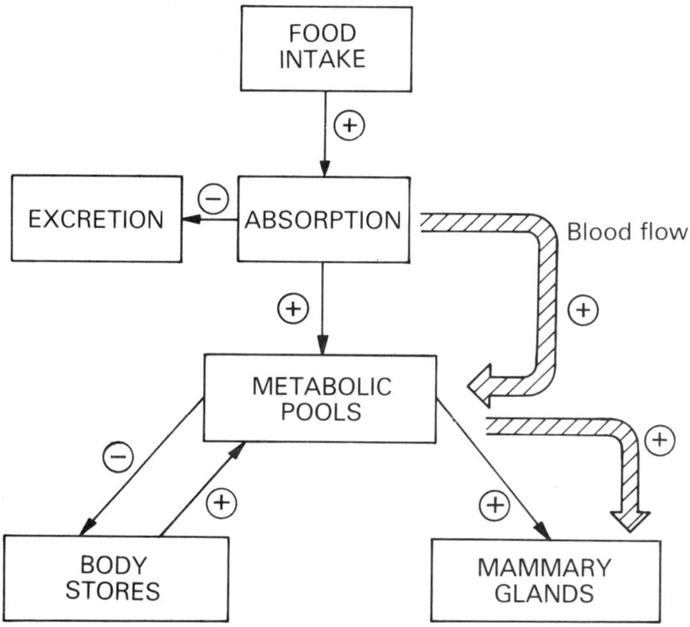

Figure 10.3. Sites of homeorhetic control during lactation. +, stimulation; −, inhibition.

Let us now consider how the postulated endocrine controls operate in relation to the major substrate needs of the glands, first by defining the changes involved and then by examining their modulation by hormones. It will probably be useful to refer to Chapter 4 at various points of the following discussion. In the next four sections we shall mainly be concerned with metabolic changes in dairy animals, changes which obviously have important implications for formulation of feed rations.

10.2.1. Partitioning of glucose

Figure 10.4 shows results of studies on goats, in which the mammary uptake of glucose was compared with its total entry rate into the circulation in four animals at six different stages. Total entry rate was determined by *isotope dilution*, in which, by intravenous infusion of ^{14}C glucose for several hours, a steady-state SRA of glucose was achieved, as the entry of isotope exactly balanced that utilized by body tissues: division of the steady-state SRA by that of the infused isotope gives a measure of total glucose entry rate. Mammary uptake was measured by AV sampling, coupled with mammary blood flow estimation (see section 4.5.1). At peak lactation, when the milk yield was 3.3 ml min^{-1} (goat J, Figure 10.4), no less than 85% of the total flux was absorbed by the mammary

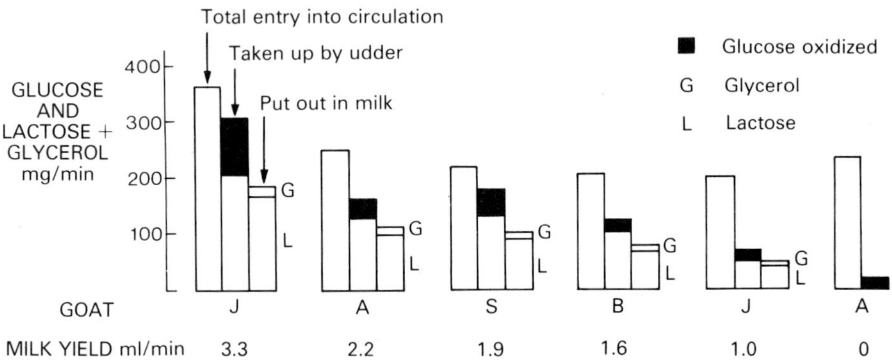

Figure 10.4. Uptake of glucose from blood plasma by goat mammary glands (centre columns) compared with its total entry rate into the circulation (columns on left) and the output of its derivatives in milk (columns on right); data for four goats.
(Source: Annison and Linzell, 1964.)

glands, so that the rest of the body had to cope with the remaining 15%. Later in lactation, when the yield had fallen to 1.0 ml min^{-1}, the mammary uptake accounted for 35% of the, much reduced, flux (goat J, Figure 10.4) and in a dry animal (goat A) the figure was only 7%.

Flux measurements only indicate the rate of supply of substrate to the circulatory system. Another important question concerns the immediate sources of that supply. It is estimated that in dairy cows the glands frequently require in excess of 3 kg of glucose daily at peak lactation. Probably not more than 1 kg is supplied by intestinal absorption, because the microbes in the rumen ferment most carbohydrate to volatile fatty acids (see section 5.2.1). The remainder must, therefore, be provided by other means. Although hepatic glycogen stores are mobilized in early lactation, the quantities available are minor. The extent to which prepartum glycogenesis in the mammary glands, demonstrated in mice (see section 9.9), occurs in other species is unknown, though it is unlikely to be quantitatively significant in cows. But certainly, in ruminants, the glucose 'deficit' is largely met by hepatic (and to a lesser extent, renal) gluconeogenesis, largely from propionate, but also to a minor degree from amino acids, lactate and glycerol.

Increased availability of glucose for milk secretion also results from its decreased use by other tissues. Thus, in a study on cows, the extent of its oxidation by the whole animal decreased from 30 to 10% between late pregnancy and early lactation. Moreover (see section 10.2.4), utilization of glucose to form the glycerol moiety of TG in adipose tissue is markedly reduced in lactation.

Before leaving discussion of glucose, it is interesting to note (see Figure 10.4) how closely the mammary uptake of glucose balances with the milk output of lactose and TG glycerol, when allowance is made for mammary glucose oxidation (derived from measurement of AV differences of CO_2 SRAs).

10.2.2. Partitioning of amino acids

Lactation exacts a rapid and formidable demand on amino acid supply. Unfortunately, we have few hard facts on the extent to which body protein reserves are mobilized, so that we have to resort to speculative calculations to gain insight into the magnitude of changes which might occur. But it seems certain that skeletal muscle protein is the major source of endogenous amino acids.

It has been estimated that a 600 kg cow has about 25 kg of protein capable of mobilization. The pattern of amino acids furnished by their proteolysis is unlikely to match mammary requirements, but, if we ignore that constraint, it can be calculated that amino acids equivalent to the protein in 800 kg of milk might, theoretically, be made available. Rarely, however, do cows lose weight at rates greater than 2 kg/day, so that an upper limit of 300 g/day of protein may be assigned to this source of supply. This is equivalent to about 10 kg of milk, i.e. about 30% of the yield in early lactation.

Data on total entry rates, equivalent to those for glucose (see Figure 10.4), are not available for amino acids, but the high mammary extractions (see section 4.5) of certain essential amino acids, for example, methionine (for which the figure may be over 80% in goats), suggest that the glands make extravagant demands on the metabolic pools.

10.2.3. Partitioning of lipids

We have noted (see section 10.2) that there is substantial mobilization of fat in early lactation. Fat is stored as TG in adipose tissue cells (*adipocytes*), the volume of which may decrease by 50% in the first 2 weeks of lactation in cows. Figure 10.5 indicates the major pathways of lipid synthesis and mobilization in the adipocyte. Lactation is attended by an increased rate of lipolysis and also by a decreased rate of lipogenesis in adipocytes, the latter being a consequence both of reduced glucose and fatty acid uptake from blood and of reduced activity of the lipogenic enzymes in the cells. By contrast, the activity of mammary lipoprotein lipase is increased during lactation (see section 9.5.1), facilitating fatty acid uptake. In high-yielding cows, body fat mobilization in the first 5 weeks of lactation is equivalent to 50% of the milk fat output during this period.

10.2.4. Partitioning of minerals

Lactation also presents a significant challenge to the body's ability to channel adequate amounts of calcium and phosphorus into milk. In cows, the efficiency

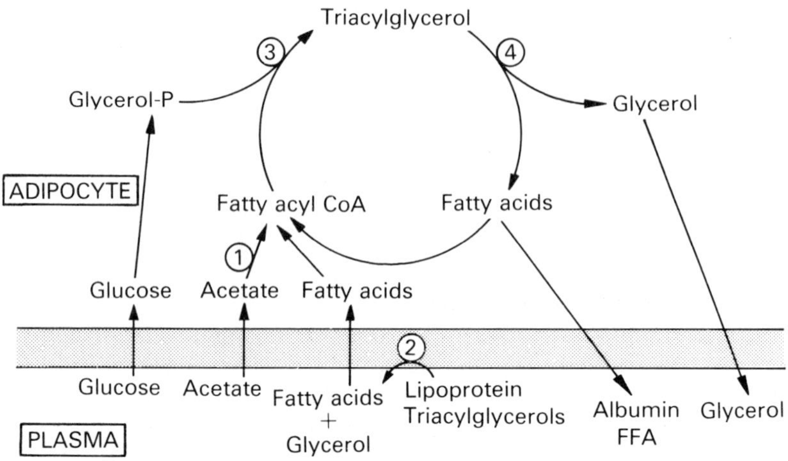

Figure 10.5. Summary of the fat synthesis–mobilization cycle in ruminant adipose tissue. 1, *de novo* fatty acid synthesis; 2, uptake of preformed fatty acids from blood; 3, fatty acid esterification; 4, lipolysis.
(Source: Bauman and Elliot, 1983.)

of gastrointestinal absorption of calcium almost doubles between late pregnancy and early lactation, but even so adequate supplies to the glands are only achieved by depletion of calcium (and phosphorus) stores in the maternal skeleton.

Intestinal absorption of Na^+, K^+, Cl^- and water also increase post partum.

10.3. Cardiovascular changes

The changes discussed above serve to increase the flux of substrates into the circulating blood, but a necessary element of partitioning is the selective distribution of these substrates to specific organs, notably the mammary glands, by means of changes in the cardiovascular system.

The cardiovascular correlates of lactation have been most extensively studied in laboratory animals. Figure 10.6 shows results of studies performed on anaesthetized rats in which, following cardiac output determination by a thermodilution method, the animals received injections of a freely diffusible tracer (^{86}Rb). Shortly after the injection procedure the animals were killed and the distribution of tracer in different organs assessed; the relative distribution of tracer thus gave a measure of the fractional distribution of cardiac output. Such studies show that not only the mammary glands but also organs concerned with nutrient absorption and metabolism, such as the gastrointestinal tract and the liver, received increased relative blood flows and gained weight during lactation. By contrast, after initial increases, the relative blood flows and weights of the skin, carcass, and notably of the uterus, declined. The specific influence of lactation in these changes is illustrated by the markedly different changes occurring in animals from which the litters were removed at birth (see Figure 10.6). (It is important to

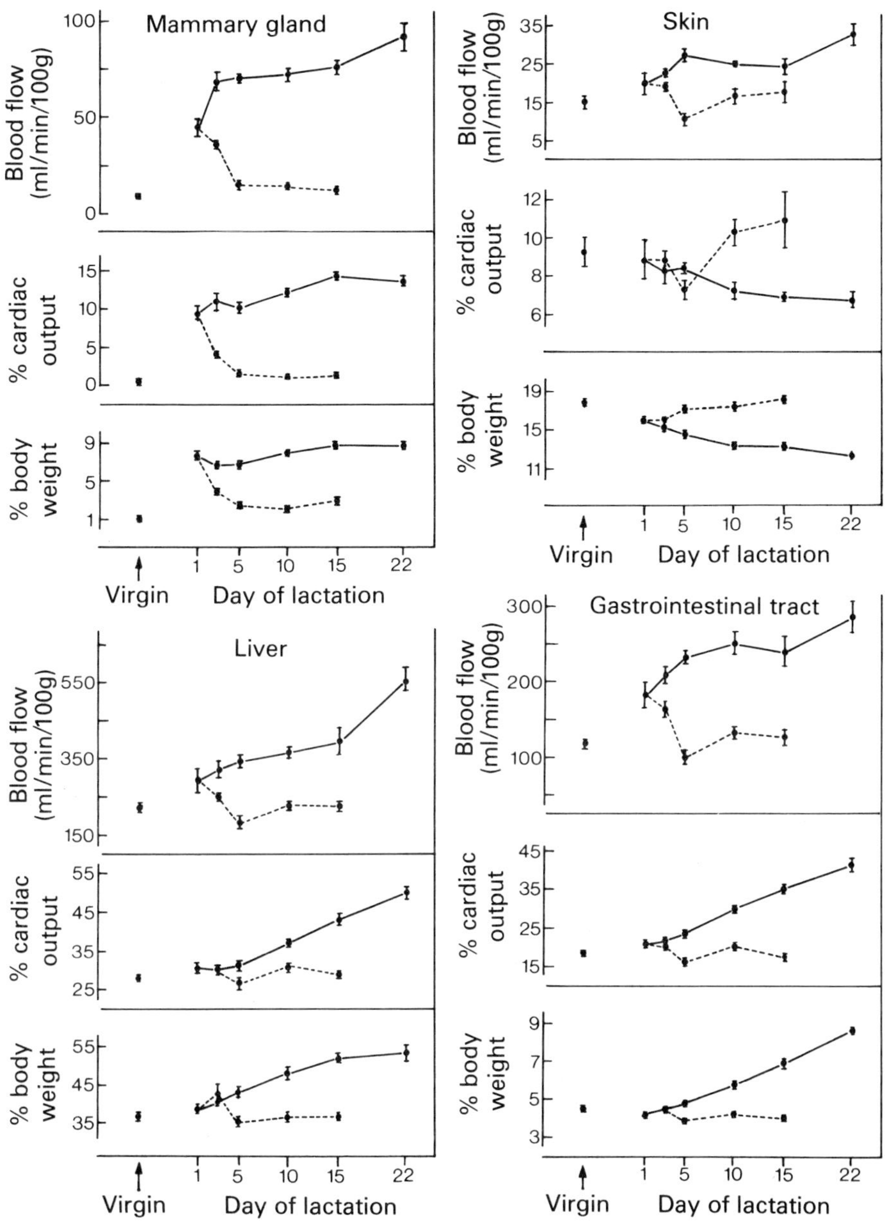

Figure 10.6. *Postpartum* changes in MBF, fractional blood flow (per cent of cardiac output), and weight of mammary glands, liver, skin and gastrointestinal tract of lactating (solid lines) and non-lactating (dashed lines) rats. Bars show standard errors of means.
(Source: Hanwell and Linzell, 1973.)

appreciate (see Fig. 10.1) that milk yield in these rats increased linearly over the period of study).

Actual blood flows (as opposed to 'relative flows') were also increased for mammary glands, liver and gastrointestinal tract, partly in consequence of elevated cardiac output and partly by vasodilatation and/or increased organ size, but the elevated cardiac output also meant that the skin maintained an increased flow during lactation, despite its reduced relative flow: the latter may be important in effecting heat transfer to the young, whose own thermoregulatory mechanisms are underdeveloped until about 18 days post partum.

Equivalent studies on large animals have not been performed, but it is known that MBF increases sharply at parturition in ruminants (see Figure 7.6) and remains higher in lactating than in non-lactating animals. Cardiac output is also elevated in lactation.

10.4 Endocrine control

We now need to consider evidence for the involvement of individual hormones in the partitioning process. This evidence is rather patchy: for example, much more is known about lipid metabolism than about protein metabolism.

10.4.1. Prolactin

The high blood concentration of this hormone exerts reciprocal effects on mammary and adipose tissue, respectively. It enhances the activity of mammary lipoprotein lipase, thus facilitating uptake of long chain fatty acids for milk synthesis, but depresses adipocyte lipase activity.

It has also been proposed that prolactin increases intestinal absorption of calcium and stimulates gut hypertrophy, but the latter claim is open to question.

10.4.2. Growth hormone

In ruminants, GH concentrations are elevated during lactation and the galactopoietic response to exogenous GH treatment (see section 7.5) suggests that it plays an important role. In part, GH may exert its effects by increasing MBF (and possibly by increasing cardiac output), but it is not known whether such effects are primary or secondary to other, well-defined, biochemical actions. GH has at least two distinct properties: lipolytic and *diabetogenic* (blood glucose elevating). During lactation, GH appears to act on adipose tissue to increase lipolysis and, probably, on hepatic cells to promote gluconeogenesis and/or glycogenolysis. It is uncertain whether GH exerts any differential effects on protein metabolism in mammary and non-mammary tissues, and indeed the loci of GH action (and the extent to which it acts via somatomedins (see section 9.4.2)) are ill-defined.

10.4.3. Insulin

During pregnancy, insulin seems to be an important regulator of lipogenesis in adipocytes, via its effect on glucose uptake (see Figure 10.5). It has been suggested that in rats progesterone induces up-regulation of adipocyte insulin receptors, while at parturition, when blood progesterone concentrations fall, the consequent fall in insulin receptors and the concomitant fall in insulin concentrations results in depression of lipogenesis. By contrast, insulin receptor numbers *increase* in mammary tissue at parturition, despite the reduced insulin concentrations. Not all species show this dependence of adipocyte insulin receptors on progesterone, but, in most, lactation is characterized by low insulin concentrations, or, perhaps more significantly, by high GH:insulin ratios.

During established lactation in rats, prolactin plays a major role in lipid metabolism, both via its effects on lipogenesis (see section 10.4.1) and by stimulating insulin receptors in mammary tissue and down-regulating them in adipose tissue. In ruminants this role is probably largely taken over, or at least augmented, by GH.

In the mammary gland, increased insulin receptors promote glucose uptake for lactose and glycerol synthesis and for oxidative purposes, as well as facilitating the uptake of certain amino acids.

10.4.4. Adrenaline and noradrenaline

The lipolytic sensitivity of adipocytes to noradrenaline, released locally from sympathetic nerve-endings, and adrenaline, reaching the cells in the blood, is markedly increased during lactation in cows. Between late pregnancy and early lactation the number of β-adrenergic receptors per cell doubles, although binding affinity is unchanged.

10.4.5. Parathyroid hormones and vitamin D

Parathyroid hormone (PTH), *calcitonin* (CT) and vitamin D are involved in almost all aspects of calcium and phosphorus metabolism, regulating intestinal absorption, and bone deposition and resorption of these elements. Vitamin D from the diet, and that produced in skin in response to light, has actions analogous to steroid hormones. Under the influence of PTH, the kidneys convert inactive vitamin D to the active form, D_3, which stimulates production of calcium-binding protein in the intestines, thus facilitating calcium absorption. Mobilization of calcium from bone is promoted by PTH and inhibited by CT, so that increased calcium supplies in lactation imply a shift of the balance in favour of PTH activity.

10.4.6. Cumulative endocrine effects

Figure 10.7 summarizes some of the endocrine effects on body metabolism during lactation. The situation is, however, much more complicated than this figure would suggest. Clearly, hormone release is due to two types of stimulation: (a) the tactile (and associated exteroceptive) stimuli of suckling (Chapter 8), and (b) the metabolic stimuli consequent on the removal from the body of large amounts of milk products, coupled with the intake of large amounts of food. Changes in hormone concentration are probably primary in the former case but may, or may not, be primary in the latter case. For example, changes in blood glucose concentration both influence, and are influenced by, changes in the GH:insulin ratio. Moreover, several other hormones and humoral agents are likely to be

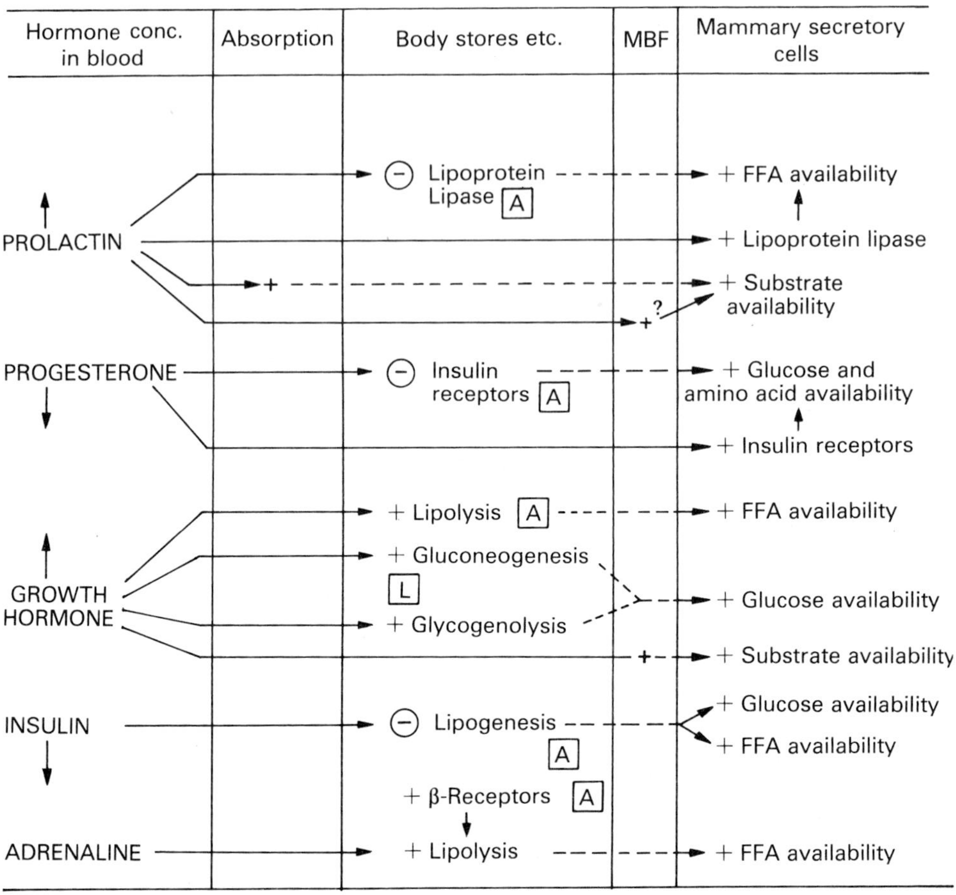

Figure 10.7. Some effects of hormones on body metabolism during lactation. +, stimulation; −, inhibition; A, adipose tissue; L, liver. Not all effects may occur in all species.

involved in homeorhetic control (such as, glucagon, somatomedins and prostaglandins), and both metabolite concentrations (for example, of glucose and acetate) and physical factors (activation of gastric stretch receptors) are probably important in the regulation of appetite.

In many respects, it is difficult to distinguish between 'cause' and 'effect' in the changes which occur. For example, does food intake increase *because* the mammary glands remove large amounts of substrates from the body pool, or is milk secretion increased as a result of a neurogenically stimulated food intake? The question is obviously not just an academic one; for example, the answer has important implications for feeding dairy animals. But the answer, in practical terms, is hedged about with lots of 'ifs' and 'buts'.

The same sort of question may be asked of the relationship between MBF and cardiac output: does the latter increase during lactation because of signals generated by the active mammary gland, or is the MBF increase secondary to elevation of cardiac output? Experimental evidence can be adduced in support of either possibility. In fact, the results of experiments on lactating rats (see section 10.3) indicate that MBF is subject to varying influences throughout the course of lactation. Figure 10.8 shows that cardiac output increased over the first 5 days of lactation (benefiting almost all organs), but between days 5 and 15 it stabilized,

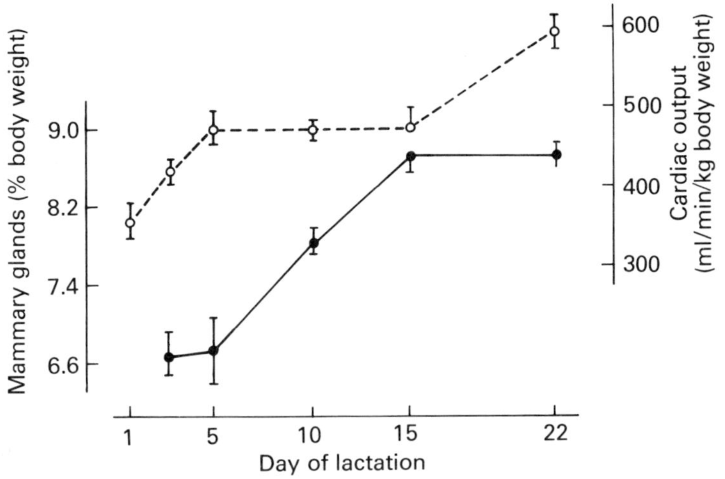

Figure 10.8. Changes in cardiac output (dashed line) and mammary weight as a percentage of body weight (solid line) during lactation in rats: these are the same animals as those for which data are presented in Figure 10.6. Bars show standard errors of means.
(Source: Hanwell and Linzell, 1973.)

increasing again between days 15 and 22. Since over this whole period milk yield increased linearly, cardiac output could not have been the sole determinant of MBF between days 5 and 15. During that period MBF continued to increase as a consequence both of growth of the glands (see Figure 10.6) and mammary vasodilatation, so that the glands received a larger proportion of the (stable) cardiac output. A possible explanation of such changes is that in both early and late lactation the demands of the young exceed the ability of the glands to satisfy them, so that the resulting vigorous suckling elicits a systemic (increased cardiac output) response. In the middle phase of lactation, however, the local control, resulting in further mammary growth and vasodilatation, is adequate to meet demand. It is known that, in rats, prolactin and GH (released in response to suckling) increase cardiac output.

10.4.7. Metabolic diseases of lactation

In high-yielding dairy animals many metabolic controls are stretched to the limit. It is hardly surprising that in some cases the limit is exceeded and serious, sometimes fatal, disease states result.

A prime example is *milk fever* (*parturient paresis* or *hypocalcaemia*), which in its advanced stages produces paralysis, coma and death. Its occurrence is almost always associated with parturition, when it is precipitated by the sudden fall in blood calcium concentrations which results when the flux of calcium into milk outstrips the body's ability to provide adequate replenishment by intestinal absorption and/or resorption from bone. A likely cause is feeding diets of too high a calcium content in late pregnancy: this depresses PTH secretion, so that when the sudden demand for calcium occurs at parturition the body is ill-equipped to make appropriate adjustments. An apparently sound strategy is to feed 'low calcium' diets prepartum.

Ketosis (*acetonaemia*) is characterized by hypoglycaemia, hyperketonaemia and sharply reduced food intake and milk yield. Occurring in early lactation, it seems to result from an imbalance between glucose supply and demand, so that acetyl CoA suffers conversion to ketones instead of being oxidized in the TCA cycle. An alternative explanation relates the disease to excessive lipid mobilization, such that the liver's ability to export adipose-derived fatty acids as lipoproteins is exceeded. Hormonal factors are likely to be implicated in either explanation. Another condition of some importance is *fat cow syndrome*, a term encompassing a range of disorders associated with obesity. Such animals have excessively fatty livers and exhibit *anorexia* (loss of appetite). Inappropriate prepartum feeding is doubtless a contributory factor.

10.5. Effects of lactation on fertility

Females have a much greater energy investment in reproduction than males, in consequence of the demands of pregnancy, lactation and maternal care. (There can be few fathers who have not felt a certain embarrassment on being congratulated on the arrival of a baby, to which their own contribution was decidedly brief and hardly burdensome.) It is thus not surprising that the regulation of fertility is principally subject to maternal physiological controls.

In many species breeding is controlled by seasonal factors: for example, annual fluctuations in day length (*photoperiod*) and temperature trigger hormonal changes which modulate oestrous activity. Periods of infertility are important in nutritional terms, both because the depletion of maternal body reserves needs to be made good and because, particularly in temperate climates, there are large seasonal variations in food availability. Lactation, however, also exerts potent contraceptive effects in some species, and as a strategy, *lactational anovulation* is more advantageous than seasonal anovulation because it may, on the one hand, persist for several years, or, on the other hand, in the event of death of the suckling infant, allow a rapid return to fertility.

In species which adopt the *r* reproductive strategy (see section 3.9.2) lactational anovulation is of only marginal importance, but in some primates it is the principal means of fertility control. In several species, seasonal and lactational effects are complementary. The ewe, for example, comes into oestrus in the autumn, and, if not mated, continues to have regular 16 day cycles until the spring. A ewe mated in August would lamb in January and, in theory, would be capable of conceiving again shortly after lambing to give birth to a second set of lambs in the following autumn. Normally, however, this does not occur because lactational anovulation takes over at a time when seasonal inhibition would not have become effective.

Marsupials show an even more impressive exploitation of the inhibitory influence of lactation. A tammar wallaby gives birth to its infant (joey) about a month after the summer solstice, but within a few hours the mother returns to oestrus, ovulates and conceives again. However, as a result of the joey's suckling activity the raised blood prolactin concentrations inhibit luteal progesterone synthesis; endometrial development is thus inhibited and the new blastocyst enters a state of *embryonic diapause.* This 'resting state' is dependent on the continued suckling stimuli of the joey for up to about 6 months—if the joey is removed from the pouch during this period the blastocyst will resume development. But after 6 months the lactational diapause is succeded by a seasonal diapause, when removal of the joey has no effect on blastocyst development. Seasonal diapause ceases to be effective at the summer solstice; the corpus luteum becomes active again and blastocyst development results in birth about 4 weeks later, rapidly followed by a new ovulation and conception. The combination of lactational and seasonal effects thus limits breeding to an annual occurrence, even though the wallaby is 'pregnant' for all but a few hours of each year.

In some higher primates, including humans, the inhibitory effects of lactation on ovarian activity are very marked, although seasonal factors may also contribute in an indirect way by influencing the nutritional status of the lactating mother. Figure 10.9 shows the time-course of return to menstruation in breastfeeding and

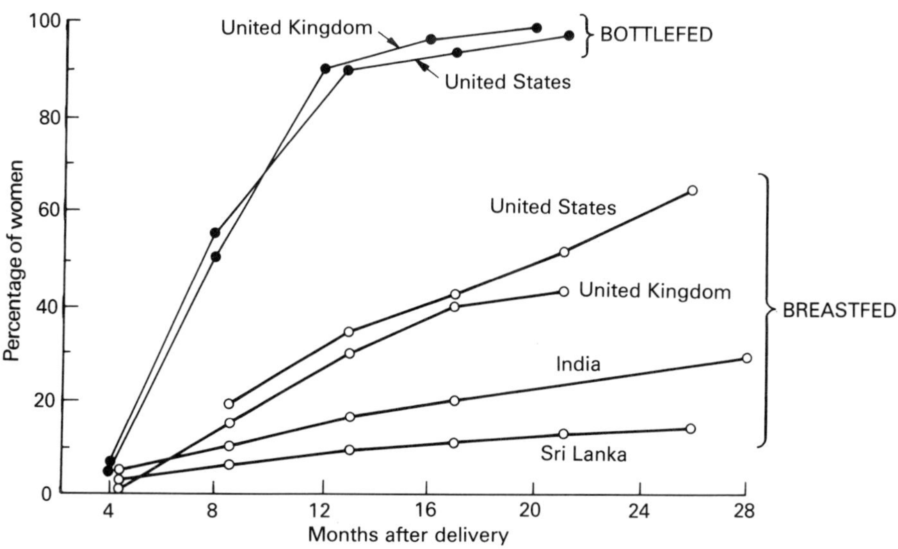

Figure 10.9. Cumulative percentage of women menstruating at different intervals after delivery, according to type of infant feeding.
(Source: Kleinman and Senanayake, 1984.)

bottlefeeding women; but note that resumption of menstrual cycles does not necessarily mean return to fertility. It is clear that breastfeeding markedly suppresses ovarian activity, although racial and/or cultural differences are also evident.

The most convincing evidence for the contraceptive potency of lactation is provided by studies on the !Kung, hunter gatherers who live in the Kalahari Desert (see section 1.8); in the absence of any form of artificial birth control, the mean birth interval is 4.1 years and the mean completed family size 4.7 children. Although differences in nutritional status may be a contributory factor, the greater effectiveness of breastfeeding in the !Kung than in other women (Figure 10.9) appears to be related to their high frequency of nursing. The mother always carries her infant in a sling so that it is able to suckle on demand. Suckling thus occurs about four times an hour, for periods of 1–2 min; frequent suckling also occurs at night. A fascinating account of the !Kung has recently been provided by Konner (1984).

R. V. Short has claimed that breastfeeding probably prevents more pregnancies, worldwide, than all other forms of contraception put together. Yet this has only begun to be recognized very recently. Thus, the book *People Populating*, a study of global population growth and birth control written in 1975 by the distinguished gynaecologist, D. Llewellyn-Jones, makes no mention of the contraceptive role of breastfeeding on any of its 368 pages.

10.5.1. Endocrine basis of the effect

There is no doubt that suckling, rather than production of milk, is the essential basis of the contraceptive effect. Thus, surgical denervation of the teats of ewes abolishes anovulation, but lactation continues normally.

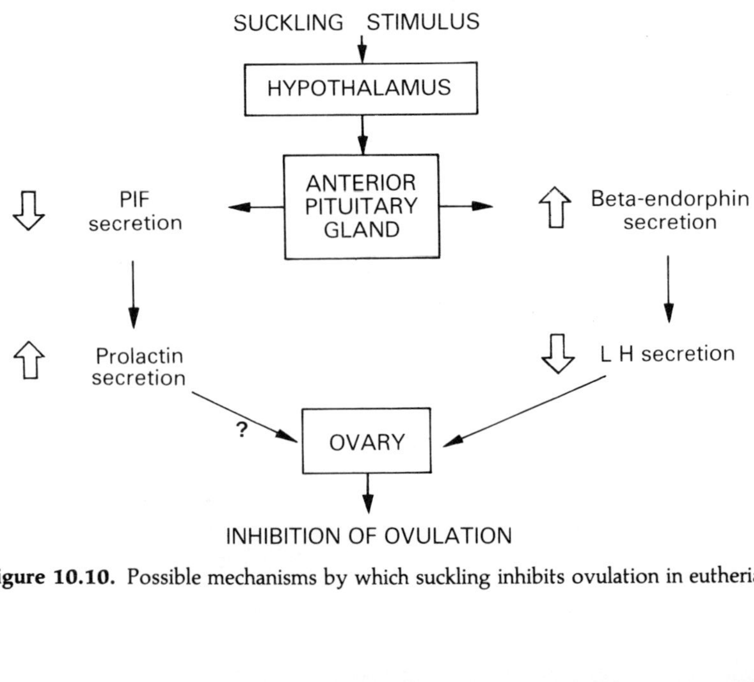

Figure 10.10. Possible mechanisms by which suckling inhibits ovulation in eutherians.

We have seen (in section 8.3.1) that afferent nerve impulses from the teats inhibit dopamine (PIF) release and result in prolactin secretion from the anterior pituitary gland. In the tammar wallaby, the resulting hyperprolactinaemia, via its suppression of luteal activity, seems to be the cause of the lactational diapause. However, this is unlikely to be the mechanism in eutherians because there is much evidence that ovarian cycles can be re-established even when blood prolactin concentrations are high. Currently, it is most widely believed that suckling increases the hypothalamic production of β-endorphin (see section 8.2.6), which suppresses hypothalamic release of GnRH (see section 8.3.6): anterior pituitary release of LH is thus decreased and ovulation inhibited. Figure 10.10 summarizes possible routes by which suckling inhibits ovulation.

Further reading

Annison, E. F. (1983). Metabolite utilization in the ruminant mammary gland, in *Biochemistry of Lactation*, pp. 399–436, ed. Mepham, T. B. (Amsterdam and New York: Elsevier.)

Bauman, D. E. and Elliot, J. M. (1983). Control of nutrient partitioning in lactating ruminants, in *Biochemistry of Lactation*. pp. 437–68, ed. Mepham, T. B. (Amsterdam and New York: Elsevier.)

Care, A. D., Bartlett, J. P. and Abdel Hafeez, H. M. (1980). Calcium and phosphate homeostasis and its relationship to the aetiology and prevention of parturient paresis, in *Digestive Physiology and Metabolism in Ruminants*, pp. 429–46, eds, Ruckebusch, Y. and Thivend, P. (Lancaster: MTP Press.)

Hanwell, A. and Linzell, J. L. (1973). The time-course of cardiovascular changes in lactation in the rat. *J. Physiol.* **233,** 93–109.

Hanwell, A. and Peaker, M. (1977). Physiological effects of lactation on the mother. *Symp. Zool. Soc. Lond.* **41,** 297–312.

Konner, M. (1984). The Tangled Wing (Harmondsworth: Penguin.)

Llewellyn-Jones, D. (1975). People Populating (London: Faber and Faber.)

Oldham, J. D. (1981). Amino acid requirements for lactation in high yielding dairy cows, in *Recent Developments in Ruminant Nutrition*, pp. 49–81, eds Haresign, W. and Cole, D. J. A. (London: Butterworths.)

Schultz, L. H. (1974). Ketosis, in *Lactation*, Vol. 2, pp. 318–53. eds Larson, B. L. and Smith, V. R. (New York and London: Academic Press.)

Short, R. V. (1984). Breast feeding. *Sci. Amer.* **250,** 23–9.

Short, R. V. (1984). Oestrous and menstrual cycles, in *Hormonal Control of Reproduction*, 2nd edition (Vol. 3 of *Reproduction in Mammals*) pp. 115–52, eds Austin, C. R. and Short, R. V. (Cambridge and London: Cambridge University Press.)

Short, R. V. (1985) Species differences in reproductive mechanisms, in *Reproductive Fitness*, 2nd edition (Vol. 4 of *Reproduction in Mammals*) pp. 24–61, eds Austin C. R. and Short, R. V. (Cambridge and London: Cambridge University Press.)

Tyndale-Biscoe, C. H., Stewart, F. and Hinds, L. A. (1984). Some factors in the initiation and control of lactation in the tammar wallaby, *Symp. Zool. Soc. Lond.* **51,** 389–401.

Vernon, R. G. and Flint D. J. (1984). Adipose tissue: metabolic adaptation during lactation, *Symp. Zool. Soc. London.* **51,** 119–45.

11 Problem-solving Strategies

> It is not enough that you should understand about applied science in order that your work may increase man's blessings. Concern for man himself and his fate must always form the chief interest of all technical endeavors, concern for the great unsolved problems of the organization of labor and the distribution of goods—in order that the creations of our mind shall be a blessing and not a curse to mankind. Never forget this in the midst of your diagrams and equations.
>
> A. Einstein (1938)

11.1. Introduction

Let us take it as axiomatic that the fruits of scientific endeavour should be used to benefit mankind. But rather than this being an incidental 'spin-off' of our pursuit of knowledge, perhaps in a world so beset with troubles, we should see science as *primarily* a 'problem-solving' exercise. Precedent suggests that we shall run no risk of stunting our intellectual development.

What, then, are the problem areas to which our existing knowledge and future enquiries of lactation might be applied? The two most obvious are breastfeeding and dairy production, but several others are of significance. Thus, lactational efficiency affects growth and survival of other animals used by man (for example, pigs and sheep) and it is an important factor in conservation programmes for rare breeds and threatened species.

We cannot, however, hope to get very far in our analysis if we confine our attention to physiological matters. The problems to be addressed lie at the centre of a complex network, in which political, social, cultural, economic and ecological, as well as biological, factors all interact. It is important to gain some idea of the nature and scale of these factors. For it is only by adopting a holistic approach to the interactions between biology and society at large that we stand any hope of successful problem-solving.

The following sections consider some of the perceived benefits and problems of the lactation process in a social context. The object is to explore how the discoveries and hypotheses of the laboratory (discussed in the earlier chapters of the book) might best be translated into actions in the real world. There will be several references to 'developed' (that is, European and North American) and 'developing' (third world) countries. This is convenient short-hand; neither category is homogeneous.

11.2. Breastfeeding

There is every reason to believe that breastfeeding was a vital phase of reproduction in all human societies until quite recent times; the practice of the !Kung (see section 10.5) was probably stereotypical. The biological rationale would seem to be twofold. In the first place, the large relative size of the human head (the morphological basis of our superior intellects) demands that birth occur at an early developmental stage; further postponement would preclude its passage through the birth canal. Postnatal nurture, breastfeeding, thus assumed a crucial role. Secondly, because the duration of this postnatal phase exceeded the capability of seasonal factors to suppress fertility (see section 10.5), it fell to the suckling stimulus to serve this role.

11.2.1. The scale of the problem

Latterly, however, things have started to go wrong. Women have increasingly sought to emancipate themselves from what to many seems a somewhat vulgar and inconvenient activity. This aversion to breastfeeding is centuries old, albeit initially on a small scale. For example, in eighteenth-century London many women from wealthier families employed 'wet nurses' to feed their babies. Since wet nurses were typically of poor health and financial status, not averse to dosing their charges with gin to silence their cries, it is hardly surprising that infant death and disease were rife. By the mid-nineteenth century in Europe, a growing appreciation of the chemical composition of milk led to the marketing of artificial foods (often based on cow's milk and wheat flour), and bottlefeeding was given impetus by the invention of the rubber teat. Initially, however, bottlefed babies thrived even *less* well than those fed by wet nurses.

By the early years of this century, great advances in sanitation, appreciation of the microbial basis of many diseases and improved formulations for artificial foods, all led to increasing safety in the practice of bottlefeeding. Babies seemed to thrive and the problem of returning fertility was solved by use of artificial contraceptives. Nowadays, many women claim that bottlefeeding is a boon, freeing them from a demanding and tiring routine.

Whether or not it is true that it has no significant disadvantages in the developed west, which in view of the benefits of breastfeeding discussed (in sections 1.7 and 1.8) is certainly questionable, there is no doubt that the widespread adoption of bottlefeeding in the very different economic and social conditions of many developing countries has had disastrous consequences.

Many women in third world countries appear to have fallen victim to the superficial attractions of a 'western style of living', in which the breast is valued more for its role in sexual arousal than in nursing. They have come to believe that use of baby food from a tin is more sophisticated (even, perhaps, more nutritious) than feeding from the breast. Some lay the blame for this situation squarely on the persuasive advertising campaigns of the baby-food manufacturers, who are said to have fostered 'commerciogenic malnutrition'. But whether or not such a charge is valid, the adverse consequences are starkly evident. Inability to afford the costs of adequate feeds (Table 11.1), deficiencies in the nutritional properties of the formula (see section 3.3.5), inadequacy of procedures used for sterilizing the feeding bottles and rapid return to fertility in the absence of breastfeeding (see Figure 10.8)—all these interact to compound the problem (see section 1.8).

Table 11.1. Cost of complete formula-feeding for an infant 2 months of age, expressed as a percentage of salaries for selected jobs in different countries. (*Source: UNICEF, 1985.*)

Country		*Percent of salary*	
	Hospital cleaner	*Ministry clerk*	*Junior staff nurse*
Burma	73	40	21
Egypt	9	10	8
England	6	5	4
Guatemala	27	12	10
Indonesia	35	51	21
Nigeria	18	16	6
Philippines (Manila)	28	24	18
Sri Lanka	63	45	43
Sweden	4	4	4
Tanzania	32	32	14
Turkey	21	21	16
Yemen	17	13	11

In *The State of the World's Children in 1985* UNICEF reported that in the third world an estimated 14 million children under the age of 4 years die each year (10 million before the age of 11 months), a figure equivalent to the entire population of under-5s in the United States (Figure 11.1). The cause of death is largely a combination of the effects of gastroenteritis and protein-energy deficiency. About 90 million of the infants who survive suffer permanent stunting and, frequently, impaired intellectual development as a result of prolonged attacks of malnutrition and infection. With only 17% of couples in developing countries

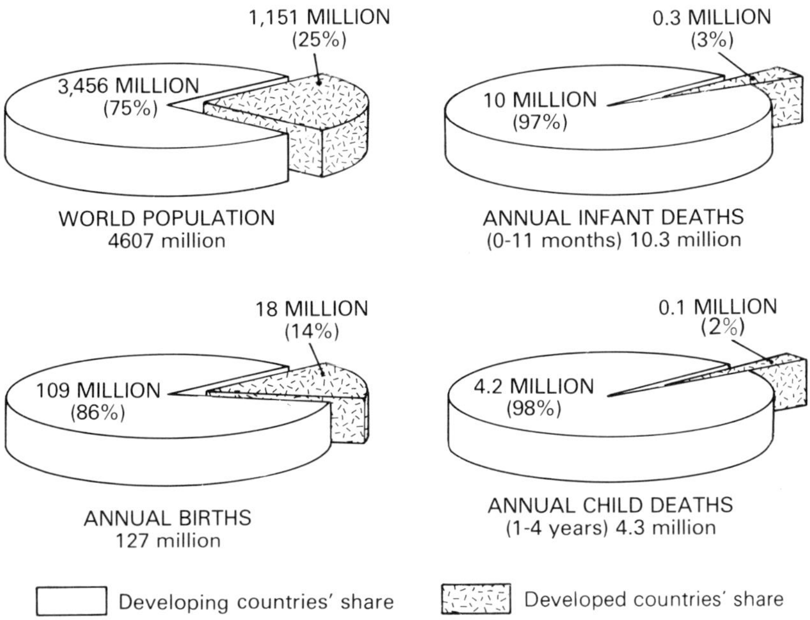

Figure 11.1. The developing world's share of population, births, and child deaths. (Source: UNICEF: 1985.)

currently using modern contraceptive methods, and in view of cultural and social impediments to their rapidly increased usage, it seems unlikely that they can be relied on to significantly reduce birth rates.

Some specific examples will illustrate the problem. In Chile, where the decline in breastfeeding has been particularly rapid (Figure 11.2), infant mortality is much higher in bottlefed babies (even when receiving supplementary breastfeeding) than in wholly breastfed babies (Figure 11.3). Infant morbidity and mortality are highly related to birth-spacing, as is evident in Bangladesh, where reduced fertility is clearly associated with reduced mortality (Figure 11.4).

There is little doubt that the large-scale readoption of breastfeeding would significantly reduce infant death and disease. Babies would receive appropriate nutrition and vital immunological protection, while far fewer would be born. It is estimated that if all Indian mothers were to breastfeed there would be 5 million fewer births annually. In developed countries there has been a resurgence of interest in breastfeeding over the last decade, chiefly among women in higher socioeconomic groups. Regrettably, no corresponding change is apparent in third world countries.

The problems discussed above are those of major quantitative significance, but there are other lactation-related difficulties of importance to many women. Some are unable to produce enough milk, in others lactation needs to be suppressed; premature (low birth weight) babies present distinctive nutritional problems. In

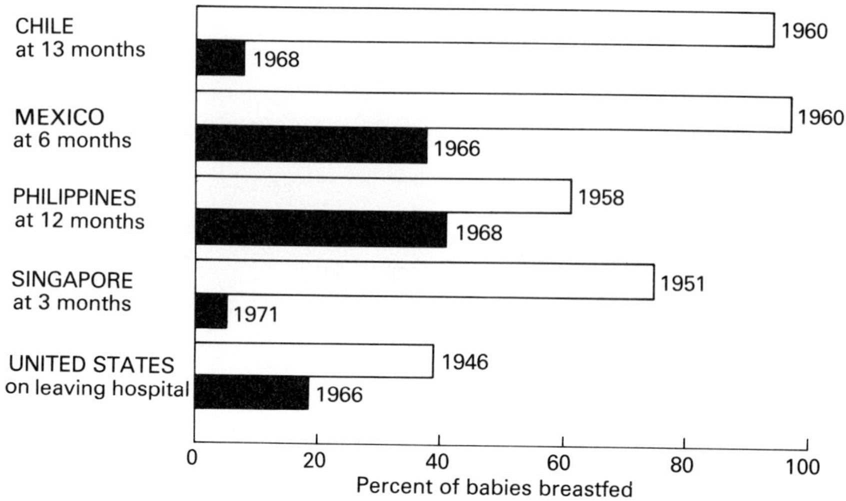

Figure 11.2. The decline in breastfeeding in some developing countries.
(Source: Ebrahim, 1978.)

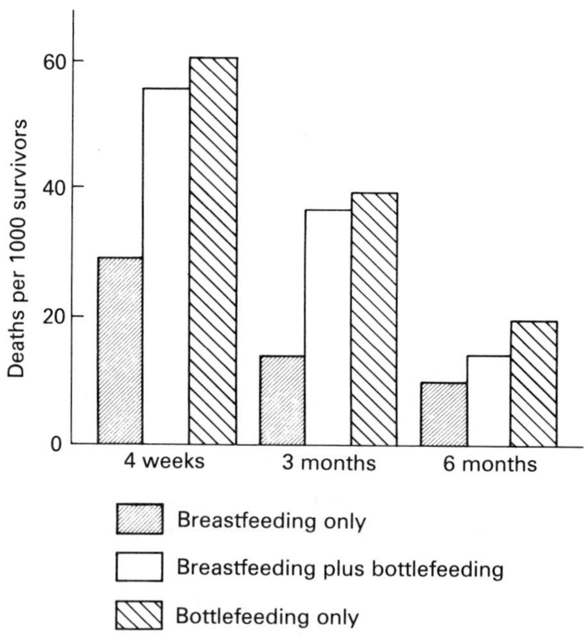

Figure 11.3. Mortality in the first year of life among infants surviving at 4 weeks, 3 months and 6 months according to type of feeding.
(Source: Plank and Milanesi, 1973.)

Spacing between births	Infant deaths /1000 births	Toddler deaths /1000 alive	Child deaths /1000 alive
	0-1st birthday	1st-2nd birthday	2nd-4th birthday
Less than 2 years	185	42	81
2-4 years	89	28	62
Over 4 years	58	10	27

Figure 11.4. Reduction of deaths in infants, toddlers and children, following increased spacing and births in Bangladesh.
(Source: Kleinman and Senanayake, 1984.)

the following sections we shall analyse the lactational process into component elements and consider how it might best be regulated to meet human needs.

11.2.2. Specific interventions in lactational control

Figure 11.5 summarizes the major advantages of breastfeeding and specific sites at which controls might be exerted (lettered arrows). The nature of these possible controls is now discussed.

A. Clearly, major factors are *education* and *motivation*. Women in both developed and developing countries need to be encouraged in, and made fully aware of the benefits of, breastfeeding. Organizations like La Leche League International play an important role in promoting breastfeeding, but the scale of the problem in developing countries would seem to merit government action. This could take the form of advertising campaigns (Figure 11.6), teaching programmes, and encouragement of older, experienced women to act as *doulas*, a term of Greek origin, reintroduced in the 1970s to describe a 'sympathetic adviser'. Use of artificial foods might be curbed by import or price controls, or by banning their use except 'on prescription'. But the need for education would seem also to apply to society as a whole: much could be achieved if shops, public buildings and work places were to make suitable provision for nursing mothers.

In recent years there have been encouraging signs that the seriousness of the problem is becoming recognized. In 1981, after a 10-year campaign against the promotion of bottlefeeding, the World Health Assembly adopted an *International Code of Marketing of Breast-Milk Substitutes*, which aims to *'promote breast-feeding*

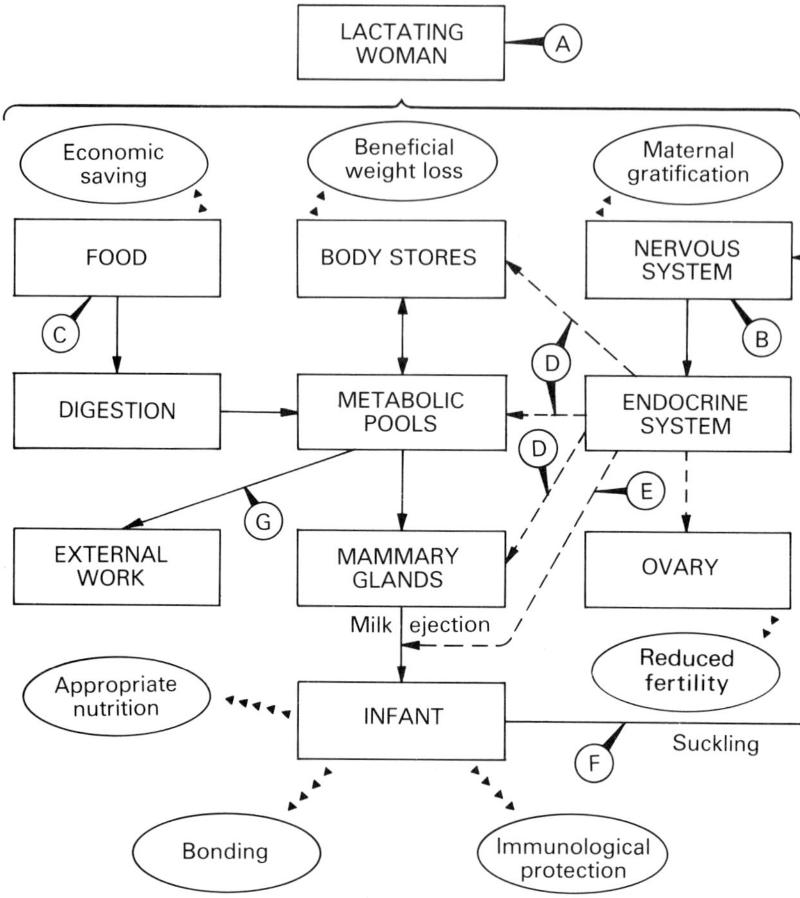

Figure 11.5. Benefits (ovals) of breastfeeding in women, and points at which the process might be controlled (lettered arrows). (For details see text.)

and (ensure) the proper use of breast-milk substitutes when these are necessary'. By 1985, 33 governments had banned the advertising of breast-milk substitutes to the public, and 17 had actively used the mass media to encourage breastfeeding. Unfortunately, the code is proving very difficult to enforce, and in just one 3-week period in 1984 an international body reported over 100 violations of the code by 22 companies in ten different countries. Clearly, the campaign to promote breastfeeding still has a great deal to achieve.

In some cases, however, for example, because of the mother's illness or death, use of an artificial food seems to be indicated. But there is also a biological option, *relactation*. This refers to the initiation of milk secretion in women who have ceased lactating, or even who have not previously lactated. Thus, an adoptive mother or relative may be able, by frequently putting the child to the breast (see section 8.3.5), to produce sufficient milk to meet its needs. The practice is

Figure 11.6. A simple, cost-free, means for the widespread promotion of breastfeeding in India.

quite common in some societies, and women have been known to nurse their grandchildren, when the mother has died.

B. Some women in both developed and developing countries are unable to produce adequate milk for reasons which seem related to pychological tension or apathy. When medical advice, or that of a *doula*, is to no avail, tranquillizers such as chlorpromazine may be effective. They act at the hypothalamus and result in increased prolactin secretion (see section 8.3.2). Tranquillizers may sometimes cause *galactorrhoea* (inappropriate milk secretion).

On occasion, it may be necessary to inhibit lactation, for example, if the mother is receiving drugs which would harm the baby if transferred in the milk. When the breast is not suckled, secretion stops in a few days (see section 7.2.6), but painful congestion may occur in this time. Administration of bromocriptine, which suppresses prolactin secretion (see section 8.3.2) is highly effective in inhibiting milk secretion and does not appear to have adverse side-effects. Alternatively, L-dopa, PIF itself (see section 8.3.1), may be administered.

C. Since lactation makes considerable demands on body reserves (see section 10.2), adequate nutrition during pregnancy and lactation are important. A well-nourished woman may gain 4 kg of fat during pregnancy. This is equivalent to about 150 MJ of energy, which is theoretically sufficient to provide the energy in milk for about 2 months (at about 2.5 MJ day^{-1}, that is about 700 ml milk day^{-1}). By contrast, a poorly fed woman may be unable to lay down any extra reserves in pregnancy (Figure 11.7), and consequently her weight may fall below that prior to pregnancy as inroads are made into her basal reserves. Despite this, the baby's requirements seem often to be met; the volume of milk may be reduced, but not, to the same degree, its solid content. As a rule, deficiencies in milk yield are best remedied by appropriate maternal nutrition, rather than by supplying the baby with supplementary food. It follows that in both developed and developing countries, additional intake of appropriate foods during lactation is important. These serve to maintain adequate supplies of nutrients on which the gland makes a high demand, for example, energy, amino acids, calcium and

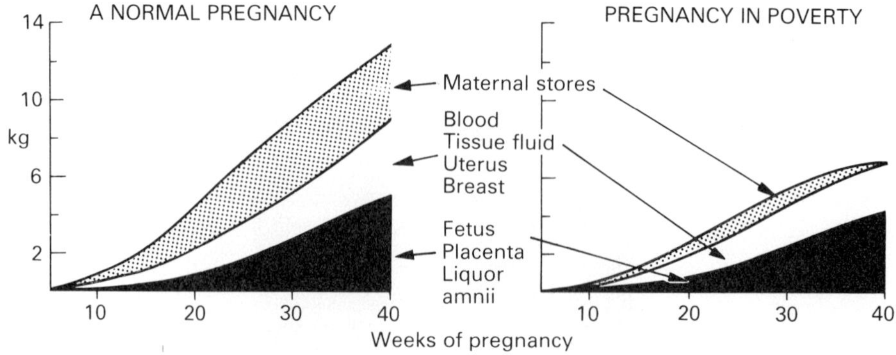

Figure 11.7. Weights of maternal and fetal constituents during a normal pregnancy and pregnancy in poverty.
(Source: Hytten and Leitch, 1971.)

vitamins. Thus, the recommended average daily intakes of energy, protein and calcium for women in Britain rise from 9.0 MJ, 54 g and 500 mg, respectively, to 10.0, 60 and 1200 during pregnancy, and to 11.5, 69 and 1200 during lactation (Department of Health and Social Security, 1979).

D. Hormonal induction of mammary growth in women seems only to have been used for cosmetic purposes. Oestrogen creams, applied locally, can produce growth, but it is chiefly of the stroma (see section 2.2.4), and disturbances to the menstrual cycle may limit their use. In any case, small breasts in non-pregnant women are not indicative of subsequent poor lactational performance: large breasts are mostly fat.

Hypogalactia (inadequate milk secretion) might be susceptible to treatment with human prolactin, but the supplies of the compound are not yet adequate for its clinical use. However, TRH (see section 8.3.1) has been shown to elicit release of prolactin from the anterior pituitary gland in women and thereby enhance yield. Since TRH is a tripeptide which is readily synthesized in the laboratory, its clinical use in some cases of hypogalactia might prove beneficial.

Oestrogens may also be used to suppress lactation, but in view of certain adverse side-effects, the use of bromocriptine (**B**, above) is now much preferred. It is perhaps not surprising that the combined oral contraceptive pill also has an inhibitory effect, and where women require additional 'cover' in the advanced stages of breastfeeding, 'gestogen-only' contraceptives, such as the 'minipill' or Depo-Provera injections, appear to be suitable.

E. Anxiety or stress may inhibit milk ejection by blocking the release of oxytocin from the posterior pituitary gland (see section 8.2.6), which, through local feedback effects (see sections 4.9.1 and 7.5.1), depresses milk secretion. The problem may be overcome by administering synthetic oxytocin, either by nasal spray or as tablets, kept in the mouth while nursing. Usage is generally only temporary, most women soon acquiring the ability to eject milk spontaneously.

F. Initiation of suckling *immediately* after birth is now recognized as an important element of successful breastfeeding. The practice at this stage has more to do with establishing a bond (see section 1.8), transferring immunoglobulins in colostrum (see section 3.3.3) and providing an apprenticeship in suckling, than with provision of nutrients. In some cultures, colostrum has been considered 'impure' (presumably because of its obviously different appearance from milk) and nursing is delayed for days or weeks post partum: this is highly detrimental.

Since the effectiveness of breastfeeding in depressing fertility is dependent on a high frequency of nursing (see section 10.5), advertising this fact would seem to be a major objective of education programmes (see **A** above).

G. Any alternative demands on energy supplies and reserves have repercussions for milk production. Moderate work has not been shown to have deleterious effects, but the hard agricultural labour, in which women in many societies traditionally engage, is likely to be detrimental to effective feeding.

11.3. Dairy milk production

Consumption of animal milk, butter and cheese has been practised by human societies for millenia. Almost invariably, docile, herbivorous animals have been used, and successfully persuaded to yield milk to the abnormal stimulus of hand-milking or machine-milking. In medieval English villages, even quite poor families owned livestock, often including a cow, and daily milk consumption may have been as much as 0.57 l (1 pint) per day, though probably much less in winter. Sheep milk was frequently used to make cheese.

The dairy industry only became important in Europe and North America after the Industrial Revolution, when the extension of railways and the introduction of pasteurization and refrigeration procedures enabled transportation of milk from the country into large towns. Previously, townspeople often drank milk from cows which were kept in wretched and unhygienic conditions within the towns. Now, in the 1980s, dairying is a major industry in the developed west, which produces 80% of the annual global milk yield of over 460 million metric tonnes.

It would be a mistake to view dairying as simply a process of milk production. It shares with other animal production systems a dependence in land, vegetation, climate and human skills. And in many countries it is just one of several uses to which animals are put. In the third world, animals are frequently the most valuable possession of peasant peoples, providing meat, hide, draught-power and materials for manufacture and fuel, as well as supplying milk. Milk, like several other animal products, has high biological value (see section 3.2) and desirable organoleptic qualities (tastiness). But often a heavy price is paid for these benefits because the energy value of the product is at best only one-third of that of the crops eaten. It follows that when dairy animals are fed crops (such as cereals) which could be consumed directly by humans, or where land use is deflected from growing such crops, the energetic efficiency of food production is reduced.

Dairy production from ruminants can often, however, be exonerated from the charge of prodigality. Because the rumen microbes effectively degrade the cellulose of plant cell walls (see section 5.2.1), ruminants can utilize herbage which humans are unable to digest. Furthermore, much land unsuitable for crop cultivation (for example, for topographical or climatic reasons) can be profitably grazed by animals such as sheep and goats. And there is much to be said for a balanced system of crop and animal production, from both ecological and economic viewpoints. Complementarity rather than competition seems to be the rule: the record harvest in India in 1983 was accompanied by the record milk yield from dairy animals.

We in the developed west should, however, be under no illusions; much of the milk (and meat) we consume has been produced by energetically wasteful processes. To compound our guilt, by disposing of surplus products to third world countries we provide a disincentive to the establishment of their home-based industries.

11.3.1. Milk and the world food problem

The number of people suffering from malnutrition together with those adjudged to be receiving inadequate diets account for about 40% of the world population of nearly 5000 million (see Figure 11.1). The problems of world food supply are very grave and exceedingly complex. But in the view of expert opinion, such as the FAO (Food and Agricultural Organization of the United Nations), the efficient attainment of increased milk production would make a major contribution to easing the problem. The advantages would not only be nutritional; there would be increased provision of employment and generation of regular cash incomes, as well as a number of benefits peculiar to animal production systems, such as availability of draught-power and fertilizer.

Milk consumption in the developing world (currently about 25 kg per person per year) is unlikely ever to attain the same importance as it does in Europe and North America (where average consumption is 300 kg per person annually) because of the prevalence of *lactose intolerance*. White Europeans and their decendents are, in fact, unusual in retaining into adulthood the ability to digest lactose, through persistence of intestinal lactase (see section 3.2). Such people seem to have originated in South-west Asia, where, 5000–8000 years ago, milk of cattle, sheep, goats and camels was consumed. It is speculated that a mutation resulting in lactase persistence had survival value in the prevailing circumstances. By contrast, in the majority of the world's peoples the activity of intestinal lactase declines in early childhood, leaving them less able to digest lactose. *Lactase deficiency* does not necessarily result in lactose intolerance, in which intestinal disorders result from bacterial fermentation of the lactose and osmotic entry of fluid into the gut lumen; but it usually restricts intake of milk in raw form.

The real value of milk to third world peoples is as a *supplement* of high quality protein (see section 3.3.5) and essential fatty acids (see section 3.4.1) to a predominantly vegetable diet. One example will suffice to make this point. A 4-year-old weighing 16 kg needs to consume about 333 g of rice daily to satisfy protein requirements: with the water absorbed in cooking this is a very large amount. But if 200 ml of milk is consumed daily, the 'rice requirement' drops to

138 g. If sago were the staple food, 8 kg daily (uncooked weight) would need to be consumed!

11.3.2. Production objectives

What, in general terms, are appropriate production objectives? Conscious of the problems of overproduction, as in the EEC, as well as of underproduction in the third world, we need to define our objectives in terms of *efficiency* and *control.* We want enough milk (and no more), as cheaply as possible—subject to all related political, economic, ecological and medical constraints. Fine control gives the ability to regulate lactation to meet demand; high efficiency maximizes the use of resources.

'Efficiency' is a term open to different interpretations. Expressed as 'output/input', its units will usually have monetary value in a capitalist economy, but may relate to 'land use' or 'time expended in husbandry' in other societies. In the west, maximal efficiency is likely to be attained by maximizing milk yield per animal, because as the yield increases, maintenance costs become a smaller fraction of total feed costs. Since total milk supply is the product (see section 4.10) of an intensity factor (yield per animal) and an extensity factor (their number), the *biological* solution to overproduction is to reduce the total number, maintaining fewer, but high-yielding, animals. But in a developing country, maximizing yields might well be detrimental to other uses of the animals. From a biological viewpoint, then, what we need to do is to identify control points in the lactation process (in the same way as when we considered breastfeeding in section 11,2). The extents to which these different means of control are exploited will be largely determined by extraneous factors.

11.3.3. Specific interventions in lactational control

Figure 11.8 summarizes benefits of dairy production systems and indicates actual or potential control points: in view of the very recent discovery of several relevant controls, some of the proposals must be regarded as conjectural, but the list is by no means exhaustive. In the following discussion these control points (lettered arrows) are considered in turn.

A. The choice of species and breed of animal to be used depends on factors such as climate, grazing land available and nature of the other uses of the animals. Most dairy animals are cattle, the two species, *Bos taurus* (European) and *Bos indicus* (zebu) accounting for 90% of global milk production. Sheep, goats and buffalo (the latter the main dairy animal in India), provide most of the remainder, but other species are important in specific locations, such as camels, ibex, llamas and yaks.

European cattle, which are very high yielding in temperate climates, do not tolerate tropical conditions readily, while, though zebu cattle are well adapted to the tropics, their yields are much lower. Third world breeding programmes are thus currently directed to introducing a proportion of temperate-breed genes into the indigenous stock, without unduly compromising their hardiness. The degree of success has been variable, the humid tropics presenting the greatest difficulties.

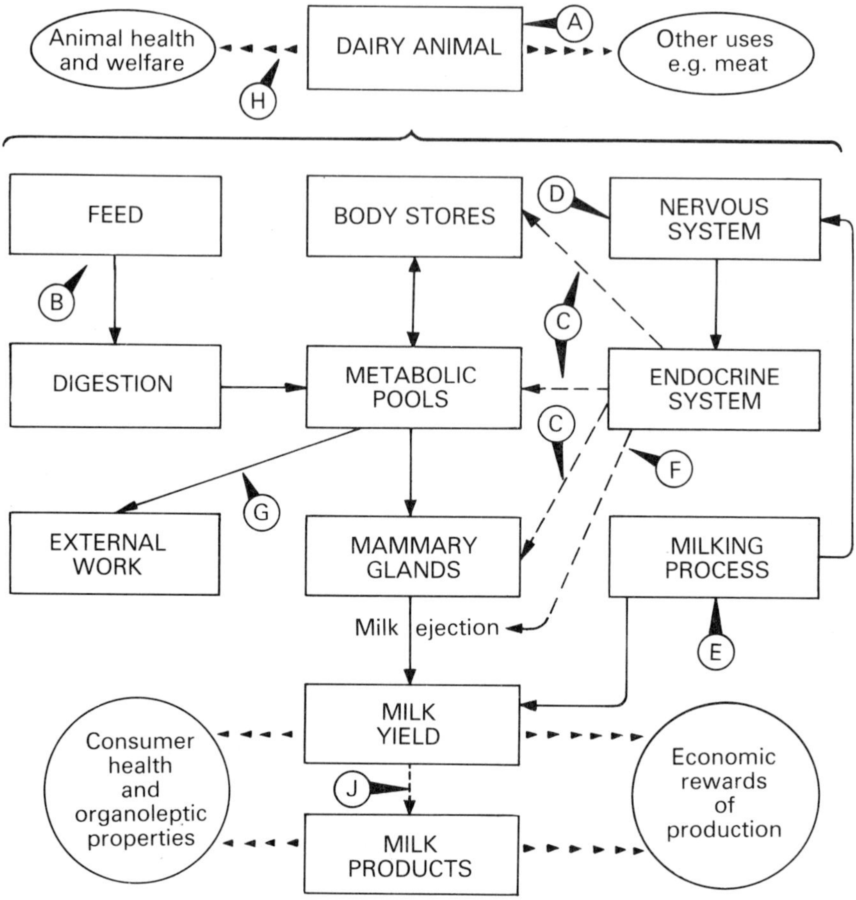

Figure 11.8. Benefits, (ovals and circles) of dairy production, and points at which the process might be controlled (lettered arrows). (For details see text.)

Artificial insemination has proved a valuable tool in these cross-breeding programmes. The recently developed technique of embryo transfer could be used to good effect, and cloning of embryos could have widespread practical applications.

In countries where land is a scarce commodity, the use of females as draught animals is often beneficial because the required draught-power can be provided by a smaller herd, while the animals can also be used for milk production.

B. Much research effort has been expended in the study of nutritional requirements of lactating cows. Only a broad outline of the principles involved is appropriate here. The range of feeding practices is very wide, so that examples of the two extremes will be considered.

In temperate climates, grazing is limited by seasonality of grass growth, and conserved forages (hay or silage) form the basis of an alternative diet. However, the forage component is generally inadequate to sustain acceptable yields so that the diet is supplemented with cereal-based concentrates, rich in protein. Their physical form (for example, whether they are ground and/or pelleted) has a marked effect on digestibility, and supplementation with non-protein nitrogen (such as urea), or prior spraying with sodium hydroxide, can improve the efficiency of feed conversion. In intensive systems, control may be exerted through changes in (a) composition, (b) volume and (c) physical form of the feed, and by changes in (d) the frequency of feed supply; all these factors are subject to (e) timing and (f) duration, in relation to the lactation cycle.

Some practical expressions of such control are: (a) exploitation of periods of good quality herbage supply; (b) variation in the concentrate:forage ratio; and (c) regulation of calcium intake in late pregnancy to avoid parturient paresis (see section 10.4.7). Possible innovations are those employing 'protected' feeds, which by escaping degradation in the rumen are more effectively absorbed into the blood. 'Protected proteins' may stimulate yield by enhancing amino acid supply, while 'protected polyunsaturated fatty acids' may be capable of producing milk with lower saturated fat content (see section 3.4.1).

At the other extreme from intensive western systems are those practised by pastoral people for whom milk is the mainstay of the diet. The lactation curve of cattle of the Boran tribe, living on the Ethiopia–Kenya border, is principally subject to control by seasonal factors, chiefly rainfall. Yields typically show two or three peaks during a single year, corresponding to the rainy seasons; in the dry seasons the Boran cattle experience extreme dehydration. Increased yield in such animals would thus seem to depend largely on agronomic factors and environmental control (see **D**, below).

C. When synthetic oestrogens became available in the 1930s experiments were performed to establish the feasibility of using them to induce mammogenesis, and hence augment yield in dairy animals, and to initiate lactation in infertile animals. Subcutaneous implantation of oestrogen 'tablets' for several weeks led to mammary growth and subsequent milk yields of up to 14 kg daily per cow. But the variability of the response, histological abnormalities in mammary structure and other troublesome side-effects precluded widespread use of the technique. Later, synthetic progesterone used in combination with oestrogens gave better results; and, latterly, short (7 day) injection treatments of the two hormones have been employed. However, growth and yield responses are still variable and inferior to those following normal development. It now seems probable that optimal growth requires placental lactogen (see section 7.3.4) or GH: a recent report suggests that GH injections are effective in stimulating development of mammary parenchyma in heifers at puberty.

Considerable success has been obtained with GH in stimulating yield in established lactation. For example, in a recent study, GH prepared by genetic engineering techniques (see section 11.4) caused a marked increase in yield in cows when injected daily for 188 days (Figure 11.9), both the peak yield and the persistency of high yield being increased. The overall increase in yield was 36.2%, much greater than that (16.5%) obtained when 'natural GH', extracted

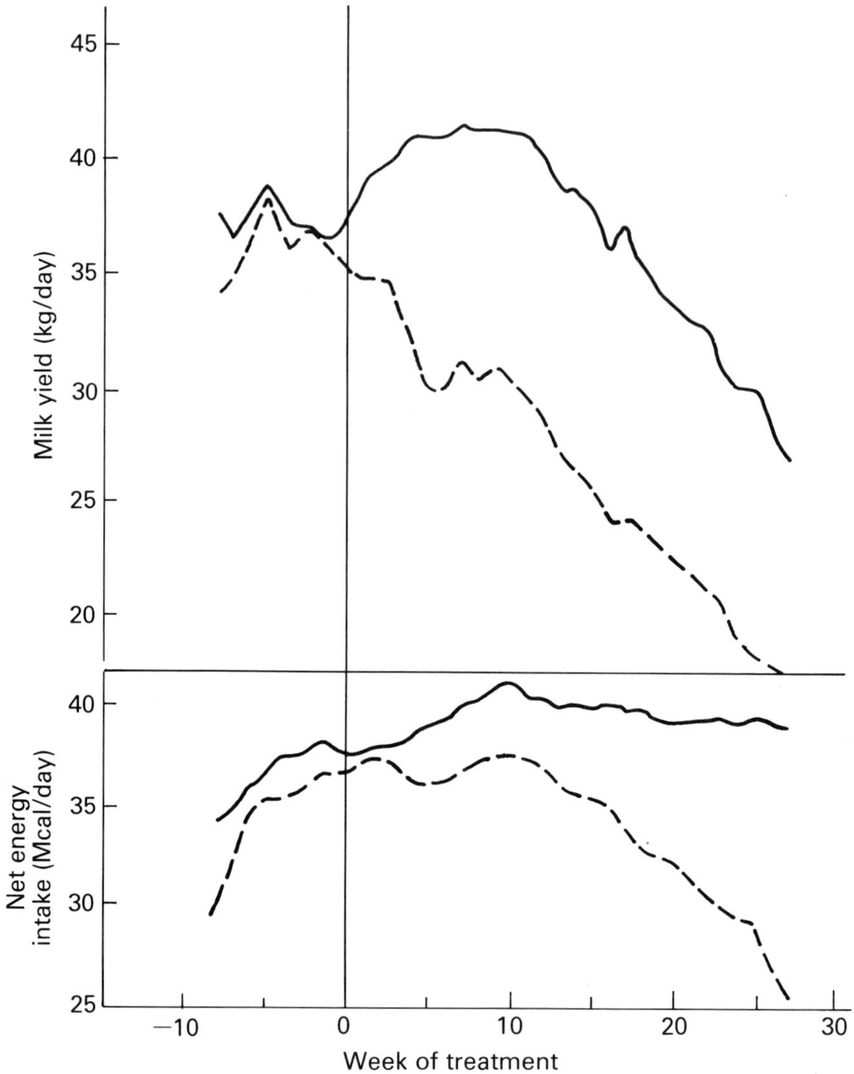

Figure 11.9. Effect of exogenous bovine GH on milk yield and dry-matter (energy) intake by lactating cows. Solid lines show values for cows treated for 27 weeks (from week 0) with 27 mg day^{-1} methionyl bovine GH, produced by genetic engineering techniques. Dashed lines show values for control (untreated) animals.
(Source: Eppard and Bauman, 1984.)

from bovine pituitaries, was used. The composition of the milk was the same as in control animals. The response seems to involve both partitioning of nutrients (see section 10.2) and increased food intake, while there is also evidence from some studies that GH increases mammary growth. There is little doubt that the galactopoietic potency of GH will soon find commercial application. An alternative means by which blood GH concentrations might be elevated would involve immunization against *somatostatin* (see Table 4.2), a natural inhibitor of GH release. Were the method to be found practicable, it would seem likely to have the benefits of ease and economy of treatment.

Thyroid hormones (see section 7.5) have also been used to increase yields in cattle, but hardly to useful effect. Hyperthyroidism is deleterious to body metabolism, while cessation of treatment leads to a rapid decline in yield.

D. Interventions at the level of the nervous system encompass many environmental stimuli as well as those of pharmacological agents.

At temperatures above 27–29 °C there is usually reduction of milk yield in cows, and of the lactose and protein content. Depressed food intake seems to be responsible, and it is possible to partially offset the effect if diets of high energy and nutrient density are fed prior to onset of heat stress. Provision of shade and showers for cooling can also be effective remedies. European breeds are the most sensitive to high temperature, but yield is usually unaffected by a fall in temperature to −4 °C.

Recent experiments with dairy cows have shown that supplemental artificial (fluorescent) lighting, to give 16 h light and 8 h dark per 24 h, stimulates yield by 6–10%, with a slight fall in fat content. The effect appears to be accounted for by the associated increase in feed intake. It is speculated that increased blood concentrations of prolactin may be responsible, but this has not been firmly established. It seems likely that installation of artificial lighting in developed countries could be highly profitable in seasons of natural short day-length. There is also some evidence that supplemental light increases the growth of mammary parenchyma in both prepubertal and postpubertal heifers.

In terms of drug therapy, bromocriptine has recently been shown to have a beneficial effect on yield in goats. This drug inhibits prolactin secretion (see section 8.3.2), but if administered only during the few days before and after parturition, the depressed yield is subsequently transformed into an *elevated* yield. Goats so treated have been shown to give 38% more milk during the whole lactation than untreated control animals. The physiological basis of the effect has not yet been fully defined, but a stimulation of glucose uptake by the mammary glands seems to be involved.

It seems possible that, in future, peptide hormone release will also be capable of control by use of opiates or their antagonists (see section 8.3.2).

E. Efficient milking is important both in maximizing current yield and in removing the inhibitor which appears to be present in milk (see section 7.5.1). Milking machines (working on a suction principle) do not closely mimic the natural suckling stimulus (see section 8.2.2) of the infant and it is conceivable that improved designs might elevate yields. It is often necessary for the calf to be present for zebu cattle to achieve efficient let-down.

Recent experiments on goats show that milking three times a day can markedly increase yields over twice-daily milking: in one series of experiments, thrice-milked goats yielded 29% more milk. Increased mammary growth may be responsible, because at the end of one experiment, thrice-milked glands were 34% heavier than controls. Since the effect has a local basis, were the identity of the inhibitor to be discovered it might be possible to develop an antibody to it and hence achieve higher yields at normal milking frequency.

A possible way of achieving higher milking rates without human labour is suggested by current developments in robotics. Cows which are given the opportunity present themselves for milking 5 to 6 times a day, and it seems likely that, with advances in micro-electronic systems for cow identification and attachment of teat cups, automatic milking of cows may soon be technically feasible.

F. Synthetic oxytocin may be used to promote let-down in animals in which the reflex is inhibited (see section 8.2.6). Its use would normally be limited to veterinary applications.

G. Alternative energy demands, for example, temperature stress (see **D**, above) or physical work, often reduce yields by partitioning nutrients away from the mammary glands. In some societies draught power is of major importance, and milk production is seen merely as a 'bonus'.

H. Animal health is of major importance, affecting both yield and animal welfare. In intensive systems, dairy animals are particularly susceptible to metabolic diseases (see section 10.4.7). Mastitis (a pathogenic bacterial infection of the udder) probably affects 50% of all cows, and although 95% of these suffer only from the subclinical form (which often goes undetected), if they are untreated (such as with antibiotics) the condition may deteriorate.

In the third world, animal diseases present formidable problems. For example, in Africa, rinderpest and foot and mouth disease are widespread.

Humanity demands that animals used by man should be protected from stress. But 'stress' is very difficult to define: anthropomorphism may be an unreliable criterion. However, milk yield appears to be peculiarly sensitive to even minor physiological disturbances (see section 8.2.6), so that depression of yield may prove to be the best objective measure of stress we can achieve.

J. Since the major objective of milk production is to satisfy human needs, it is appropriate to consider the treatment of milk following milking.

Milk is a very perishable commodity, its high nutrient content making it prone to spoilage by micro-organisms: at best, it can become unfit for consumption; at worst, a lethal carrier of disease. The commonest form of processing of liquid milk is pasteurization, in which milk is heated to, for example, 73 °C for 15 s and then cooled rapidly. Ultra-high temperature (UHT) treatment consists of heating homogenized milk to about 140 °C for at least 1 s, this permitting storage for months in appropriate containers. Traditionally, however, longer-term benefits from milk have involved manufacture of cheeses and fermentation to various products such as yoghurt, kefir and kumiss.

The different products of milk play an important role in adding variety to the diet, although processing may lead to some loss of nutritive value, for example, in respect of vitamins B and C, and in the chemical score (see section 3.3.5) of the protein. But in at least one respect, there is also a clear nutritional advantage. The benefits of yoghurt in the diet of lactose-intolerant people (see section 11.3.1) have only recently been substantiated. Yoghurt is produced by adding to milk

Lactobacillus bulgaricus and *Streptococcus thermophilus*, both of which are now known to contain bacterial lactase. This enzyme is able to survive passage through the stomach and to substitute for a lack of intestinal lactase. Thus, when lactose-intolerant people consume yoghurt, not only is its lactose content less than that of milk, but residual lactose is largely hydrolysed by the bacterial lactases, which become activated in the small intestine. Another recently introduced remedy is to add to the milk a tablet containing lactase, of fungal origin, prior to drinking.

Two properties of camel milk make it particularly suitable for people in drought-stricken areas. Firstly, its water content increases when drinking water is scarce. Secondly, it is remarkably resistant to souring: stored at 4 °C it does not sour for 3 months, and even at 30 °C it only sours after 7 days.

11.3.4. A phasic approach to lactational control

The control points (**A–H**) discussed have been assumed to operate in a stable network of interrelated elements (see Figure 11.8). But the lactation cycle consists of several phases, in which the relative importance of these elements varies greatly. Thus an alternative, more sophisticated, strategy for problem-solving might focus on modulation of these different phases (Figure 11.10a), i.e. (1) first pregnancy, (2) early lactation, (3) peak lactation, (4) declining lactation, (5) dry period, (2) early lactation ... etc.

Since it has proved possible to enhance yields, by specific intervention, to values above the normal peak (see section 11.3.3**C**), we can set a theoretical target yield 'T', which ideally we should like to maintain for as long as possible

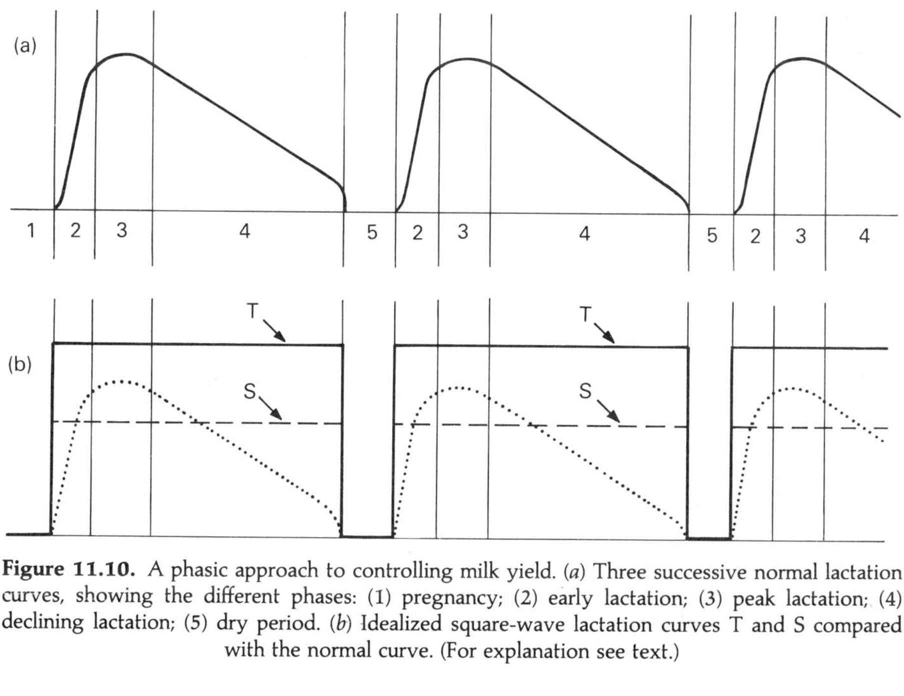

Figure 11.10. A phasic approach to controlling milk yield. (*a*) Three successive normal lactation curves, showing the different phases: (1) pregnancy; (2) early lactation; (3) peak lactation; (4) declining lactation; (5) dry period. (*b*) Idealized square-wave lactation curves T and S compared with the normal curve. (For explanation see text.)

during the period of lactation—that is, having a square-wave form (Figure 11.10b). Different strategies might well be required to achieve T during the different phases listed above, because factors like MBF (see section 10.3), hormone receptor numbers (see section 9.2) and food intake (see section 10.2) change appreciably during the course of lactation. Even a submaximal yield 'S', below current peak yields, might however be considered preferable to the normal lactation curve (Figure 11.10b).

But an even more radical approach is possible. It is known that animals (such as goats) may be milked continuously for years without re-mating, the yield fluctuating with season, but showing an overall decline. It is not unlikely that this fall in yield is partly due to cell loss, but since cell division and lactation do not appear to be incompatible (see section 7.2.5), a means might be devised to stimulate growth sufficient to compensate for this steady regression. It would thus be possible to obtain milk continously, rather than intermittently. Sustaining milk yield in excess of normal values (N) by stimulation of continuous renewal of mammary cells (Figure 11.11b) would doubtless qualify to be termed a 'utopian' yield (U) (Figure 11.11a), but even a sustained yield (R) below current peak yield might prove to be preferable to the normal pattern of lactation cycles. The notions advanced are, of course, highly speculative.

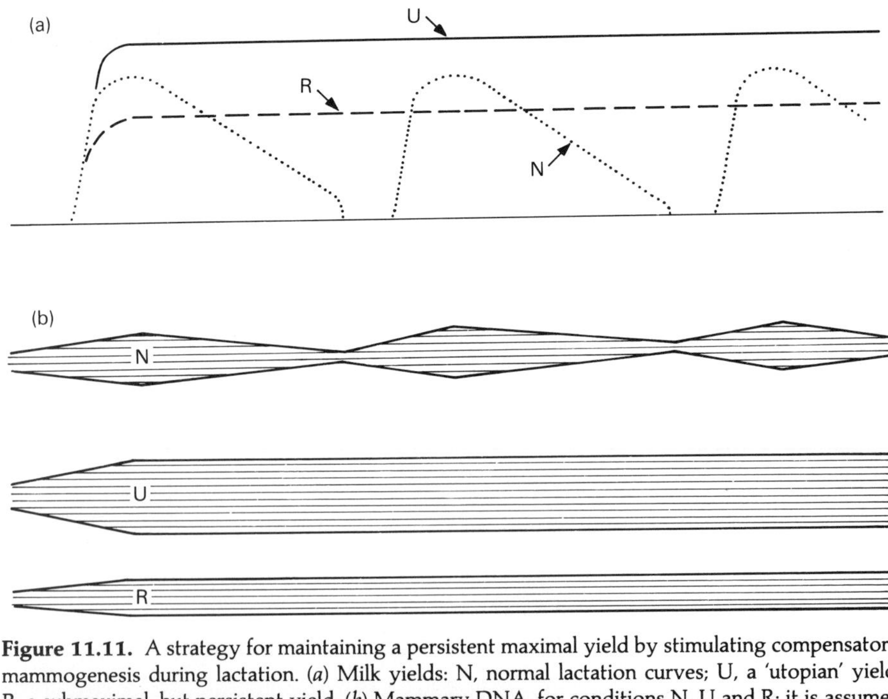

Figure 11.11. A strategy for maintaining a persistent maximal yield by stimulating compensatory mammogenesis during lactation. (*a*) Milk yields: N, normal lactation curves; U, a 'utopian' yield; R, a submaximal, but persistent yield. (*b*) Mammary DNA_t for conditions N, U and R: it is assumed that mammary DNA_t continues to increase until peak yield in a normal lactation. (For explanation, see text.)

11.3.5. Compositional manipulation

The main emphasis of the preceding sections has been that of optimalizing milk yield in dairy animals. But animal milk is not itself an ideal dietary for human consumption. We have noted that its saturated fatty acid content (see section 3.4.1) and, for many people (see section 11.3.1), lactose content place restrictions on its use. It follows that a sensible problem-solving strategy should address such matters: what we really want is the maximum capability to manipulate milk composition.

There are two approaches to solving this problem: (a) modifying the treatment of the animal to cause it to secrete milk of more acceptable composition, and (b) modifying the composition of the milk after its secretion. The latter is, of course, widely practised (see section 11.3.3J) to protect its nutritional value from spoilage, but it can also take the form of removing the fat to produce a skimmed milk more acceptable to many consumers. The more radical, and perhaps ultimately more efficient, solution is, however, approach (a).

Surprisingly little research effort has been directed to physiological modification of milk composition; the investigated use of 'protected polyunsaturated fatty acid' feeds (see section 11.3.3**B**) is a rare example of this strategy. Nevertheless, public perceptions of the adverse consequences of high animal fat intake are beginning to affect the price paid to farmers for milk. In Britain, in May 1985, the MMB (Milk Marketing Board) price for milk protein content exceeded, for the first time, that paid for fat content—a clear response to public demand for lower-fat milk. There are indications that the trend will continue.

A conceptual model for tackling this problem might well be based on the scheme shown in Figure 11.10. Instead of milk yield curves, one can envisage a range of curves, such as 'saturated fatty acid yield' or 'calcium yield'. What would then be required amounts to a definition of the rate-limiting factors controlling secretion of individual milk constituents in the different phases of lactation, with the object of stimulating or suppressing them, according to the nutritional needs of the consumer.

This brings us to the question of research priorities. An important distinction between the 'benefits' of breastfeeding and of dairy production (see Figures 11.5 and 11.8), respectively, is that while it is reasonable to expect that *all* of the former will accrue, those for dairy production are in competition. A hard-worked draught animal will not give much milk; the profits of the dairy industry can only be achieved at the expense of the consumer.

Most agricultural and food research in developed countries has been based on the implicit assumption that food production must be viewed solely as a commercial enterprise. It is, at least, arguable that a more appropriate objective is improvement of the quality of food in relation to the health and welfare of the community as a whole. The strength of this claim would appear to be overwhelming in third world countries.

11.4. Genetic engineering

In *The Gene Business* (1983), E. Yoxen quotes an example of genetic engineering: 'Bacteria do not have mammary glands, nor do they secrete milk proteins, but scientists can make them do that by introducing the necessary genetic instructions.' Words meant to startle are sometimes misleading. Synthetic milk is not yet on the doorstep, nor even a twinkle in milkman's eye.

But, in theory, the technology now exists to manufacture, by recombinant DNA techniques, not just cow's milk—but any milk we choose. Even, the much vilified 'baby-food' could be made to match human milk to perfection and this would have clear advantages in cases of medical need: production of an ideal milk for premature babies would also be a justifiable application. An engineered milk for adults would doubtless omit lactose and most of the fat. Already genetic engineering is having an impact on the lactation process. The GH used in experiments discussed above (see section 11.3.3**C**) was produced by bacteria.

This new-found potential presents society with all the old familiar problems—those of priority, propriety and property: of deciding how important it is, how it should be used, and for whose benefit. If such decisions are made solely on the basis of commercial profit, the third world will inevitably lose out once again. Is it too much to hope that the international scientific community might be able to impress on national political leaders, not only the inhumanity of hunger, but also the folly of the growing disparity between rich and poor? Scientific truth knows no national frontiers: nor will the vengeance of the dispossessed.

11.5. Milk as a diagnostic medium

11.5.1. Assessment of reproductive status

Milk, quite fortuitously, provides a valuable means of assessing the reproductive status of cows. Following successful insemination there is a rise in blood progesterone concentrations (see section 7.3.4), which is reflected in other body fluids, including milk. Using the radioimmunoassay facilities available at a central laboratory, a farmer is thus able to obtain an indication of whether insemination has been successful, about 3 weeks later. However, because of embryo losses the test is only about 80% accurate. Confirmation can be provided by assay of the milk *oestrone sulphate* concentration. This hormone is produced by interaction of the conceptus and endometrium, and the increased concentration which can be detected 105 days or more after insemination serves as a reliable index of successful pregnancy. There is also a sharp rise in the concentration of oestrone sulphate in mammary secretions a day or so prior to parturition, so that assays can provide a precise warning of the impending event. More generally, assay of steroid hormones in milk is a useful technique for diagnosing the causes of infertility in certain cows.

The routine use of diagnostic procedures on farms requires that they be simple, accurate and trouble-free. Very recently a novel approach to oestrous-detection has been investigated with the aim of satisfying these criteria: measurement of the temperature of the milk by means of a thermocouple inserted into the milking apparatus. The principle involved is based on the hypothesis that milk temperature is positively correlated with body core-temperature, which is known to increase transiently at ovulation. Such studies are currently at a very early stage and it is uncertain whether they will prove capable of practical application.

11.5.2. Detection of mastitis

Changes in the electrical conductivity of milk, which result from operation of the paracellular route of ion transfer (see section 4.9.3), can be used to detect subclinical mastitis. A suitable conductivity meter can be incorporated into a milking machine. Changes occurring in the four quarters of the cow's udder are compared by an automated system, providing a reliable index of mastitis.

11.6. Lactation as a subject of research and teaching

It is not uncommon to encounter the view that science has ceased to be a productive enterprise. 'Don't we know enough already?' and 'Look where it's got us!' are shibboleths of this modern counterculture. And when one considers that the power of applied science (exemplified by instant mass communication via television and by the potential for instant mass annihilation from nuclear weapons) has appeared impotent in solving the major problems of hunger and disease, it is not difficult to understand such a reaction. However, a deficiency of this attitude is that it overlooks the perspective of 'time'. Today's and, as importantly, tomorrow's problems need more research not less; the only way out of the mess is to apply all available resources to problem-solving. According to this view (developed by N. Maxwell in *From Knowledge to Wisdom*, 1984), social problems ought to be absorbed into the domain of scientific enquiry, in order the more effectively to facilitate their solution. It is an approach the author has attempted to pursue in this chapter.

The reader can hardly have failed to appreciate not only how much we seem to know about lactation, but also how much we are aware we do not know. The speculative nature of many of the perceived 'controls' in Figures 11.5 and 11.8, emphasizes the need for further research. But the significance of much research on the physiology and biochemistry of lactation is not limited to its application to problems of milk secretion, important as they are. For biochemists and endocrinologists, the mammary gland is often a highly suitable system in which to study mechanisms of metabolic control. Its high rate of synthetic activity, the ease with which uncontaminated samples of its secretory product can be obtained, the accessibility of the organ and its blood supply in many species, its fine sensitivity to the influence of hormones, which elicit marked and dramatic effects—all these constitute for many scientists an ideal system for investigation. Moreover, the gland lends itself, in one or other common species, to use in a remarkably wide range of physiological methods, for example, isolated organ perfusion, transplantation, arteriovenous sampling, intracisternal infusion and *in vitro* culture. Because of the absence of 'turnover' of secretory product in the duct system, it is also highly suitable for isotope studies designed to establish precursor–product relationships.

As a teaching model, lactation may be used to exploit such attributes to the full. In a way which is perhaps more justified than for most other systems, the connection between subcellular events, at one extreme, and global problems, at the other, may be made evident. The study of lactation can hardly fail to satisfy the, justifiably, growing demand from many students for 'relevance'.

And because of the profusion of connections with other disciplines, scientific or otherwise, it is a subject which presents a continual challenge to the teacher. But this is not without its reward. In the apposite words of the Hassidic proverb, attributed to the Rabbi Akiba (*c* 130 AD), which are believed to refer to the pupil–teacher relationship, 'more than the calf wishes to suck does the cow yearn to suckle'.

Further reading

Breastfeeding

Bond, J. T., Filer, L. J., Leveille,G. A., Thomson, A. M. and Weil, W. B. (1981). (eds) *Infant and Child Feeding* (New York: Academic Press.)

Brown, R. E. (1978) Relactation with reference to application in developing countries. *Clin. Pediatr.* **17,** 33–7.

CIBA Foundation (1976). *Breast Feeding and the Mother.* Ciba Foundation Symposium **45**. (Amsterdam: Elsevier.)

Department of Health and Social Security (1979). Recommended daily amounts of food energy and nutrients for groups of people in the United Kingdom. (London: H.M.S.O.)

Ebrahim, G. J. (1978). *Breast Feeding: The Biological Option.* (London: Macmillan.)

Hartmann, P. E., Rattigan, S., Prosser, C. G., Saint, L. and Arthur, P. G. (1984). Human lactation: back to nature. *Symp. Zool. Soc. London.* **51,** 337–68.

Hesling, E. and King, F. S. (1982). *Breast Feeding in Practice.* (Oxford: Oxford University Press.)

Howie, P. W. and McNeilly, A. S. (1982). Effect of breast-feeding patterns on human birth intervals. *J. Reprod. Fert.* **65,** 545–57.

Jelliffe, D. B. and Jelliffe E. F. P. (1978). *Human milk in the modern world.* (Oxford and New York: Oxford University Press.)

Kleinman, R. L. and Senanayake, P. (1984) (eds) *Breast Feeding: Fertility and Contraception.* (International Planned Parenthood Federation.) (London)

McClaren, D. S. (1985). Health and nutritional implications of artificial milk feeding of young children in developing countries, in *Milk Production in Developing Countries.* pp. 2–14, ed. Smith, A. J. (Edinburgh: Centre for Tropical Veterinary Medicine).

Neville, M. C. and Neifert, M. R. (1983). (eds) *Lactation: Physiology, Nutrition and Breast Feeding.* (New York and London: Plenum.)

Packard, V. S. (1982). *Human Milk and Child Feeding.* (New York: Academic Press.)

Popkin, B. M., Bilsborrow, R. E. and Akin, J. S. (1982) Breast-feeding patterns in low-income countries, *Science* **218,** 1088–93.

Räihä, N. C. R. (1985). Protein quantity and quality in infant feeding, in *Milk Proteins '84*, pp. 142–53, eds Galesloot, T. E and Tinbergen, B. J. (Wageningen: Pudoc.)

Raphael, D. (1976). *The Tender Gift: Breast Feeding.* (New York: Schocken Books.)

Rowland, M. G. M., Paul, A. A. and Whitehead, R. G. (1981). Lactation and infant nutrition. *Brit. Med. Bull.* **37,** 77–82.

Short, R. V. (1984). Breast feeding. *Sci. Amer.* **250,** 23–9.

Vorherr, H. (1974). *The Breast.* (New York and London: Academic Press).

Vorherr, H. (1978). Human lactation and breast feeding, in *Lactation*, Vol. 4, pp. 182-280, ed. Larson, B. L. (New York and London: Academic Press.)

United Nations Children's Fund (UNICEF) (1984, 1985). *The State of the World's Children 1984. The State of the World's Children 1985.* (Oxford: Oxford University Press.)

Wade, N. (1974) Bottle-feeding: adverse effects of a Western technology. *Science* **184**, 45–8.

Wickes, I. (1953). A history of infant feeding. *Arch. Dis. Child.* **28**, 151–8; 232–40; 332–40; 416–22; 495–502.

Dairy production

Agricultural Research Council (1984). *Feeding Strategies for Dairy Cows.* (London: ARC.)

Bachmann, M. R. (1985). Appropriate methods of milk processing in developing countries, in *Milk Production in Developing Countries*, pp. 99–105, ed Smith, A. J. (Edinburgh: Centre for Tropical Veterinary Medicine.)

Bauman, D. E., Eppard, P. J., Degeeter, M. J. and Lanza, G. M. (1985). Responses of high-producing dairy cows to long-term treatment with pituitary somatotropin and recombinant somatotropin. *J. Dairy Sci.* **68**, 1352–62.

Beede, D. K., Collier, R. J., Wilcox, C. J. and Thatcher, W. W. (1985). Effects of tropical and subtropical climates on milk yield and composition (short term effects), in *Milk Production in Developing Countries*, pp. 322–47, ed. Smith, A. J. (Edinburgh: Centre for Tropical Veterinary Medicine.)

Cowie, A. T., Forsyth, I. A. and Hart, I. C. (1980). *Hormonal Control of Lactation.* (Berlin, Heidelberg and New York: Springer-Verlag.)

Food and Agriculture Organization (FAO) (1983). *The State of Food and Agriculture: World Review, Livestock Production, 1982.* (Rome: Food and Agriculture Organization of the United Nations.)

Flatt, W. P. and Moe, P. W. (1974) Nutritional requirements for lactation, in *Lactation*, Vol. 3, pp. 311–47, eds larson, B. L. and Smith, V. R. (New York and London: Academic Press.)

Haresign, W. and Cole, D. J. A. (1981). (eds) *Recent Developments in Ruminant Nutrition.* (London: Butterworths.)

Henderson, A. J. and Peaker, M. (1984). Feed-back control of milk secretion in the goat by a chemical in milk. *J. Physiol.* **351**, 39–45.

Hodges, J. (1985). Strategies for dairy cattle improvement in developing countries, in *Milk Production in Developing Countries*, pp. 198–217, ed. Smith A. J. (Edinburgh: Centre for Tropical Veterinary Medicine.).

Knight, C. H. (1984). Mammary growth and development strategies of animals and investigators. *Symp. Zool. Soc. Lond.* **51**, 147–170.

Kon, S. K. (1972) *Milk and Milk Products in Human Nutrition*, 2nd edition (revised). (Rome: Food and Agriculture Organization of the United Nations.)

Mathers, J. C., Pearson, R. A., Sneddon, J. C., Matthewman, R. W. and Smith. A. J. (1985). The use of draught cows in agricultural systems with particular reference to their nutritional needs, in *Milk Production in Developing Countries*, pp. 476–96, ed. Smith, A. J. (Edinburgh: Centre for Tropical Veterinary Medicine.)

Maule Walker, F. M. (1984). The effect of the hormone prolactin on milk yield, in *Dairy Science Now* August, pp. 6–7. (London: Agricultural and Food Research Council.)

Newbould, F. H. S. (1974). Microbial diseases of the mammary gland, in *Lactation*, Vol 2, pp. 269–316, eds Larson, B. L. and Smith, V. R. (New York and London: Academic Press.)

Nicholson, M. J. L. (1985). Pastoralism and milk production, in *Milk Production in Developing Countries*, pp. 424–36, ed. Smith, A. J. (Edinburgh: Centre for Tropical Veterinary Medicine.)

Ørskov, E. R. and Dolberg, F. (1985). Recent advances in ruminant nutrition and their relevance to milk production in developing countries, in *Milk Production in Developing Countries*, pp. 177–92, ed. Smith, A. J. (Edinburgh: Centre for Tropical Veterinary Medicine.)

Paige, D. M. (1982). Lactose digestion and milk intolerance in human populations, in *Animal Products in Human Nutrition*, pp. 373–83, eds Bietz, D. C. and Hansen, R. G. (New York and London: Academic Press.)

Sastry, N. S. R. (1985). Prospects of milk production from buffalo, in *Milk Production in Developing Countries*, pp. 497–517, ed. Smith, A. J. (Edinburgh: Centre for Tropical Veterinary Medicine.)

Touchberry, R. W. (1974). Environmental and genetic factors in the development and maintenance of lactation, in *Lactation*, Vol. 3, pp. 349–82, eds Larson, B. L. and Smith, V. R. (New York and London: Academic Press.)

Tucker, H. A. (1985). Photoperiodic influences on milk production in dairy cows, in *Recent Advances in Animal Nutrition*, pp. 211–21, eds, Haresign, W. and Cole, D. J. A. (London: Butterworths.)

Urban, E. (1985). Yogurt: a magic potion for the lactose intolerant, *Gastroenterology* **88**, 209–10.

Wheelock, V. and Fallows, S. (1985) Implications of the COMA report on 'Diet and Cardiovascular Disease' for British agriculture. (Bradford: Food Policy Research, University of Bradford.)

Yagil, R., Saran, A and Etzion, Z. (1984). Camels' milk: for drinking only? *Comp. Biochem. Physiol.* **78A**, 263–66.

Young, V. R. and Pellet, P. L. (1985). Milk proteins with reference to human needs, and World food supply and nutritional situation, in *Milk Proteins '84*, pp. 18–46, eds Galesoot, T. E and Tinbergen, B. J. (Wageningen: Pudoc.)

Yousef, M. K. (1982). (ed) *Animal Production in the Tropics*. (New York: Praeger).

Miscellaneous

Bonner, J. T. (1980). *The World's People and the World's Food Supply*. (Burlington, North Carolina: Carolina Biological Supply Co.).

Bowman, J. C. (1977). *Animals for Man*. (London: Edward Arnold.)

Caliendo, M. A. (1979) *Nutrition and the World Food Crisis*. (New York: Macmillan Publishing Co. Inc.; London: Collier Macmillan Publishers.)

Drummond, J. C and Wilbraham, A. (1958) *The Englishman's Food: Five Centuries of English Diet* (revised edition). (London: Jonathon Cape.)

Dumont, R. and Cohen, N. (1980). *The Growth of Hunger*. (London: Marion Boyars.)

Heap, R. B. and Holdsworth, R. J. (1981). Hormone assays in reproduction and fertility. *Br. Vet. J.* **137**, 561–571.

Holmes, J. H. G. (1983). Animal production and the world food situation, in *Domestication, Conservation and Use of Animal Resources*, pp. 255–284, eds Peel, L. and Tribe, D. E. (Amsterdam: Elsevier.)

Linzell, J. C., Peaker, M. and Rowell, J. G. (1974). Electrical conductivity of foremilk for detecting subclinical mastitis in cows. *J. Agric. Sci. Camb.* **83**, 309–325.

Maxwell, N. (1984). *From Knowledge to Wisdom*. (Oxford: Blackwell.)

Yoxen, E. (1983). *The Gene Business*. (London and Sydney: Pan Books.)

Index

For many purposes the List of Contents (pp. iii–ix) may be the most useful reference guide. Fig. 1 of the Preface (p. xi) summarizes the book's contents diagrammatically.